# The NSTA Reader's Guide to A FRAMEWORK FOR K–12 SCIENCE EDUCATION

## Practices, Crosscutting Concepts, and Core Ideas

**Second Edition**

# The NSTA Reader's Guide to A FRAMEWORK FOR K–12 SCIENCE EDUCATION

## Practices, Crosscutting Concepts, and Core Ideas

**Second Edition**

By Harold Pratt

Arlington, Virginia

1840 Wilson Blvd., Arlington, VA 22201
*www.nsta.org/store*
For customer service inquiries, please call 800-277-5300.


16 15 14 13 4 3 2 1



**Cataloging-in-Publication data is available from the Library of Congress.**

ISBN 978-1-938946-19-6
eISBN 978-1-936959-77-8

# The NSTA Reader's Guide to *A Framework for K–12 Science Education*

*by Harold Pratt*

## PART IV: Understanding *A Framework for K–12 Science Education*: Top Science Educators Offer Insight

# Background

In 2012, the National Research Council published *A Framework for K–12 Science Education: Practices, Crosscutting Concepts, and Core Ideas* (*Framework*)*, which identifies key scientific ideas and practices all students should learn by the end of high school. The *Framework* serves as the foundation for new K–12 science education standards that will replace those developed in the 1990s, including the *National Science Education Standards* (*NSES*) and *Benchmarks for Science Literacy* (*Benchmarks*).

A state-led effort to develop the new science standards—called *Next Generation Science Standards* (*NGSS*)—has been managed by Achieve Inc. The process involved science experts, science teachers, and other science education partners. The first draft of the *NGSS* appeared in May 2012, and the final version was released in April 2013. NSTA recommends that the science education community fully examine the *Framework* and explore in-depth the concepts and ideas on which the new standards are built.

* National Research Council (NRC). 2012. *A framework for K–12 science education: Practices, crosscutting concepts, and core ideas.* Washington, DC: National Academies Press.

# Using This Guide

This guide is intended for many audiences—including science teachers, science supervisors, curriculum developers, administrators, and other stakeholders in science education—to help them better understand and effectively implement the new standards. As the introduction to *A Framework for K–12 Science Education* (*Framework*) states, "the framework is intended as a guide to standards developers as well as to curriculum designers, assessment developers, state and district science administrators, professionals responsible for science-teacher education, and science educators working in informal settings" (pp. 2 and 4). Teachers play a key leadership role in each of these functions and will benefit from a deep understanding of the *Framework* as a stand-alone document and as a guide to the use of the *Next Generation Science Standards.*

To make the best use of this guide, the reader should have a copy of the *Framework* in hand for reference. The *Framework*, and many other National Research Council reports noted in this document, can be downloaded free of charge from the National Academies Press at *www.nap.edu.* This guide is designed to facilitate the study of the *Framework*, not replace reading it. For each chapter of the *Framework*, the guide provides

1. an overview;
2. an analysis of what is similar to and what is different from previous standards and benchmarks; and
3. a suggested action for science teachers, science supervisors, and other science educators to support understanding of the *Framework* and anticipate its impact on classrooms, schools, and districts.

## Contents of the *Framework*

The overview is not intended to be an exhaustive summary of the *Framework* chapter, but rather a brief synopsis of the key idea(s). The second section—an analysis of what is new and different—is much more effective if the reader of this guide has a copy of the *National Science Education Standards* and *Benchmarks for Science Literacy* in hand or is reasonably familiar with these documents. Much of our analysis is based on comparisons with these two important documents that were published in the mid-1990s. Other documents will also be referenced to provide additional background and reading. The third section—suggested action—contains recommendations for activities for individuals, small teams, or larger groups to explore and learn about the ideas and concepts in the *Framework*. While some will find the overview and analysis sections most insightful, others will appreciate the suggested actions and use them as guides for possible professional development ideas.

# Summary

The executive summary states the purpose and overarching goal of *A Framework for K–12 Science Education* (*Framework*) is to "ensure that by the end of 12th grade, *all* students have some appreciation of the beauty and wonder of science; possess sufficient knowledge of science and engineering to engage in public discussions on related issues; are careful consumers of scientific and technological information related to their everyday lives; are able to continue to learn about science outside school; and have the skills to enter careers of their choice, including (but not limited to) careers in science, engineering, and technology" (p. 1).

The *Framework* recommends that science education be built around three major dimensions, which are provided in the sidebar (Box S.1, p. 3)

The intent is that the *Next Generation Science Standards* should integrate these three dimensions. The early sections of the *Framework* do not communicate this intent, but it becomes clear in Chapter 9, "Integrating the Three Dimensions," and in the Chapter 12 recommendations to Achieve Inc. The early chapters are instead designed to provide an understanding of each separate dimension.

## The Three Dimensions of the *Framework*

**1. Scientific and Engineering Practices**

- Asking questions (for science) and defining problems (for engineering)
- Developing and using models
- Planning and carrying out investigations
- Analyzing and interpreting data
- Using mathematics and computational thinking
- Constructing explanations (for science) and designing solutions (for engineering)
- Engaging in argument from evidence
- Obtaining, evaluating, and communicating information

**2. Crosscutting Concepts**

- Patterns
- Cause and effect: Mechanism and explanation
- Scale, proportion, and quantity
- Systems and system models
- Energy and matter: Flows, cycles, and conservation
- Structure and function
- Stability and change

**3. Disciplinary Core Ideas**

Physical Sciences

PS1: Matter and its interactions
PS2: Motion and stability: Forces and interactions
PS3: Energy
PS4: Waves and their applications in technologies for information transfer

Life Sciences

LS1: From molecules to organisms: Structures and processes
LS2: Ecosystems: Interactions, energy, and dynamics
LS3: Heredity: Inheritance and variation of traits
LS4: Biological evolution: Unity and diversity

Earth and Space Sciences

ESS1: Earth's place in the universe
ESS2: Earth's systems
ESS3: Earth and human activity

Engineering, Technology, and the Applications of Science

ETS1: Engineering design
ETS2: Links among engineering, technology, science, and society

*Source:* NRC 2012, p. 3

# **PART I:**
# A Vision for K–12 Science Education

## Chapter 1
# Introduction: A New Conceptual Framework

## Overview

The best description of the general vision of *A Framework for K–12 Science Education* (*Framework*) is provided on pages 8–9:

> *The framework is designed to help realize a vision for education in the sciences and engineering in which students, over multiple years of school, actively engage in science and engineering practices and apply crosscutting concepts to deepen their understanding of the core ideas in these fields. The learning experiences provided for students should engage them with fundamental questions about the world and with how scientists have investigated and found answers to those questions. Throughout the K–12 grades, students should have the opportunity to carry out scientific investigations and engineering design projects related to the disciplinary core ideas.*
>
> *By the end of the 12th grade, students should have gained sufficient knowledge of the practices, crosscutting concepts, and core ideas of science and engineering to engage in public discussions on science-related issues, to be critical consumers of scientific information related to their everyday lives, and to continue to learn about science throughout their lives. They should come to appreciate that science and the current scientific understanding of the world are the result of many hundreds of years of creative human endeavor. It is especially important to note that the above goals are for all students, not just those who pursue careers in science, engineering, or technology or those who continue on to higher education.*

Also from the introduction (p. 10):

> *The committee's vision takes into account two major goals for K–12 science education: (1) educating all students in science and engineering and (2) providing the foundational knowledge for those who will become the scientists, engineers, technologists, and technicians of the future. The framework principally concerns itself with the first task—what all students should know in preparation for their individual lives and for their roles as citizens in this technology-rich and scientifically complex world.*

The chapter discusses the rationale for including engineering and technology and for the exclusion of the social, behavioral, and economic sciences. It also includes a brief description of how the *Framework* was developed by the National Research Council (NRC) committee.

## Analysis

The stated vision reinforces what has been well accepted as the vision for science education for the past two decades and is clearly articulated in the *National Science Education Standards* (*NSES*) and *Benchmarks for Science Literacy* (*Benchmarks*).

A major difference you will notice is that the *Framework* introduces and defines engineering and technology and outlines the reasons for their inclusion in the *Next Generation Science Standards* (*NGSS*).

What's also new is that to achieve the goal, the *Framework* moves science education toward a more coherent vision by (1) building on "the notion of learning as a developmental progression"; (2) focusing "on a limited number of core ideas in science and engineering"; and (3) emphasizing "that learning about science and engineering involves integration of the knowledge of scientific explanations (i.e., content knowledge) and the practices needed to engage in scientific inquiry and engineering design" (pp. 10–11).

## Suggested Action

Compare the *Framework*'s vision and overarching goals for science education to those of your state, school, or district. What differences do you find? A review and possible update by your curriculum committees might be in order because the nature of the vision and goals stated in the *Framework* will undoubtedly appear in the *NGSS*. Note the increased emphasis on how students learn science in the means or goals of how the vision will be achieved. This will be discussed in more detail in the next chapter.

## Chapter 2

# Guiding Assumptions and Organization of the *Framework*

## Overview

The *Framework* defines several guiding principles about the nature of learning science that underlie the structure and content of the *Framework*. Below is a summary of these principles, adapted from pages 24 through 29.

**Children are born investigators:** In the early years of life, children engage in and develop their own ideas about the physical, biological, and social worlds and how they work and, thus, can engage in scientific and engineering practices beginning in the early grades.

**Focusing on core ideas and practices:** The *Framework* is focused on a limited set of core ideas to allow for deep exploration of important concepts and time for students to develop meaningful understanding of these concepts through practice and reflection. The core ideas are an organizing structure to support acquiring new knowledge over time and to help students build capacity to develop a more flexible and coherent understanding of science.

**Understanding develops over time:** Student understanding of scientific ideas matures over time—across years rather than in weeks or months—and instructional supports and experiences are needed to sustain students' progress.

**Science and engineering require both knowledge and practice:** Science is not just a body of knowledge that reflects current understanding of the world; it is also a set of practices used to establish, extend, and refine that knowledge. Both elements—knowledge and practice—are essential.

**Connecting to students' interests and experiences:** For students to develop a sustained attraction to science and for them to appreciate the many ways in which it is pertinent to their daily lives, classroom learning experiences in science need to connect with students' own interests and experiences.

**Promoting equity:** All students should be provided with equitable opportunities to learn science and become engaged in science and engineering practices—with access to quality space, equipment, and teachers to support and motivate that learning and engagement, and with adequate time spent on science.

The balance of the chapter outlines the structure of the *Framework* and its three dimensions—scientific and engineering practices, crosscutting concepts, and disciplinary core ideas—and their progressions across grades K–12.

## Analysis

The introduction to this chapter lists the NRC publications *Taking Science to School* (Duschl, Schweingruber, and Shouse 2007), *America's Lab Report* (Singer, Hilton, and Schweingruber 2006), *Learning Science in Informal Environments* (Bell et al. 2009), *Systems for State Science Assessments* (Wilson and Bertenthal 2006), and *Engineering in K–12 Education* (Katehi, Pearson, and Feders 2009) that served as background for the writers of the *Framework*. These reports are based on research from the 15 years following the publication of the *NSES* and *Benchmarks* and represents an evolving knowledge of how students learn science and the nature of curriculum and instruction that will facilitate the learning. That increased level of knowledge about how students learn is reflected in the guiding principles outlined on the previous page.

## Suggested Action

Obtain copies of the publications cited in this chapter and form study or discussion groups to become familiar with the research synthesized in them and their view of how students learn science. Explore how the research and ideas have changed since the publication of the *NSES* and *Benchmarks* and how they are reflected in the *Framework*. One of the best places to begin is with *How People Learn: Brain, Mind, Experience, and School* (Bransford, Brown, and Cocking 2000). This seminal work is easy to read, contains research on the broad topic of how learning occurs, and has a chapter with examples on how students learn science, mathematics, and history. In addition, a recent report that has had significant influence on the *Framework* is *Taking Science to School* (Duschl, Schweingruber, and Shouse 2007). This report provides the background for the *Framework*'s guiding principles and helps explain the evolution from the language of inquiry to practices.

# PART II:
# Dimensions of the *Framework*

## Chapter 3
# Dimension 1: Scientific and Engineering Practices

## Overview

This chapter continues and strengthens one of the principal goals of science education, "to engage in scientific inquiry" and "reason in a scientific context" (p. 41). In doing so, it explains the transition or evolution from inquiry to practices and discusses the reasons why practices are considered to be an improvement over the previous approaches.

The change is described as an improvement in three ways:

- "It minimizes the tendency to reduce scientific practice to a single set of procedures" (p. 43).
- By emphasizing the plural practices, it avoids the mistaken idea that there is one scientific method.
- It provides a clearer definition of the elements of inquiry than previously offered.

*A Framework for K–12 Science Education* (*Framework*) identifies eight practices that are essential elements of a K–12 science and engineering curriculum and describes the competencies for each practice. They are identified and described in "Scientific and Engineering Practices" below.

### Scientific and Engineering Practices

**Asking Questions and Defining Problems**

| | |
|---|---|
| A basic practice of the **scientist** is the ability to formulate empirically answerable questions about phenomena to establish what is already known, and to determine what questions have yet to be satisfactorily answered. | **Engineering** begins with a problem that needs to be solved, such as "How can we reduce the nation's dependence on fossil fuels?" or "What can be done to reduce a particular disease?" or "How can we improve the fuel efficiency of automobiles?" |

**Developing and Using Models**

| | |
|---|---|
| **Science** often involves the construction and use of models and simulations to help develop explanations about natural phenomena. | **Engineering** makes use of models and simulations to analyze systems to identify flaws that might occur or to test possible solutions to a new problem. |

**Planning and Carrying Out Investigations**

| | |
|---|---|
| A major practice of **scientists** is planning and carrying out systematic scientific investigations that require identifying variables and clarifying what counts as data. | **Engineering** investigations are conducted to gain data essential for specifying criteria or parameters and to test proposed designs. |

**Analyzing and Interpreting Data**

| | |
|---|---|
| **Scientific** investigations produce data that must be analyzed to derive meaning. Scientists use a range of tools to identify significant features and patterns in the data. | **Engineering** investigations include analysis of data collected in the tests of designs. This allows comparison of different solutions and determines how well each meets specific design criteria. |

| Using Mathematics, Information and Computer Technology, and Computational Thinking | |
|---|---|
| In **science,** mathematics and computation are fundamental tools for representing physical variables and their relationships. | In **engineering,** mathematical and computational representations of established relationships and principles are an integral part of the design process. |
| **Constructing Explanations and Designing Solutions** | |
| The goal of **science** is the construction of theories that provide explanatory accounts of the material world. | The goal of **engineering** design is a systematic approach to solving engineering problems that is based on scientific knowledge and models of the material world. |
| **Engaging in Argument From Evidence** | |
| In **science,** reasoning and argument are essential for clarifying strengths and weaknesses of a line of evidence and for identifying the best explanation for a natural phenomenon. | In **engineering,** reasoning and argument are essential for finding the best solution to a problem. Engineers collaborate with their peers throughout the design process. |
| **Obtaining, Evaluating, and Communicating Information** | |
| **Science** cannot advance if scientists are unable to communicate their findings clearly and persuasively or learn about the findings of others. | **Engineering** cannot produce new or improved technologies if the advantages of their designs are not communicated clearly and persuasively. |

For each practice, the *Framework* includes a comparison of how the practice is seen in science and engineering, a list of student goals to achieve by grade 12, and a discussion of the progression to reach those goals from the early grades through grade 12. Box 3-2 (p. 50), "Distinguishing Practices in Science From Those in Engineering," provides a very useful three-page table.

The *Framework* repeatedly emphasizes that practices are not taught in isolation but are an essential part of content instruction. Consider this quote from page 2 (emphasis added): "the committee concludes that K–12 science and engineering education should focus on a limited number of disciplinary core ideas and crosscutting concepts, be designed so that students continually build on and revise their knowledge and abilities over multiple years, and support the *integration of such knowledge and abilities with the practices* needed to engage in scientific inquiry and engineering design."

## Analysis

The notion of moving from the language of inquiry to that of practices, and the inclusion of engineering practices, will most likely require an adjustment or paradigm shift for many science educators. For the experienced teacher or science educator who is familiar with the inquiry standards in *National Science Education Standards* (*NSES*) and has helped students meet them through the use of "inquiries," the practices will not seem that foreign. The added details and explanations of the practices will be an advantage to many users.

The parallel discussion of each practice in both science and engineering does not imply that the two should be taught or learned at the same time, but rather intends to point out the similarities and differences among the practices in both disciplines. In some sense, the science practices have emerged from *Taking Science to School* (Duschl, Schweingruber, and Shouse 2007) and *Ready, Set, Science!* (Michaels, Shouse, and Schweingruber 2008), both of which provide a review of the research on how students learn science and how that can be used in the creation of teaching materials and classroom instruction. The *Framework* builds on this research and has identified engineering practices as a parallel discussion.

In past years, science practices have not received the same emphasis that has been placed on content knowledge, nor has the integration of content and inquiry been achieved to any great extent. The *Next Generation Science Standards* (*NGSS*) most certainly will include an equal and integrated emphasis. Consider this quote from page 26: "Science is not just a body of knowledge that reflects current understanding of the world; it is also a set of practices used to establish, extend, and refine that knowledge. Both elements—knowledge and practice—are essential." The integration of practices with the content will improve students' understanding of the concepts and purposes of science and will avoid the teaching and learning of the competencies of inquiry in isolation.

## Suggested Action

The shift for most science educators in this area will be the movement from the language and standards of inquiry in the *NSES* to the language of practices and becoming familiar with the engineering practices. To gain a better understanding of engineering, obtain *Engineering in K–12 Education: Understanding the Status and Improving the Prospects* (Katehi, Pearson, and Feders 2009) and *Standards for K–12 Engineering Education?* (NRC 2010b), two of the many documents referenced at the end of this *Framework* chapter, and use them as resources for study and discussion. Both can be downloaded for free from the National Academies Press at *www.nap.edu.*

Compare the practices of inquiry in your instruction, instructional materials, and assessment to those in the *Framework* to see what may need to be added or spelled out in more detail. Notice the progression of the goals for each practice. Check your grade level for the practices against those in the *Framework.* To what extent are they integrated with the content in your curriculum? Since the *NGSS* integrates the three dimensions (see Chapter 9), beginning to review how practices of inquiry are integrated in your existing instruction—as well as how they are aligned and progress from level to level—will enhance your ability to use the new standards.

## Chapter 4
# Dimension 2: Crosscutting Concepts

## Overview

This chapter outlines the second dimension of the *Framework*, seven crosscutting concepts that have great value across the sciences and in engineering and that are considered fundamental to understanding these disciplines:

1. Patterns
2. Cause and effect: Mechanism and explanation
3. Scale, proportion, and quantity
4. Systems and system models
5. Energy and matter: Flows, cycles, and conservation
6. Structure and function
7. Stability and change

## Analysis

Readers familiar with the *NSES* and *Benchmarks* will recognize that the *Framework*'s crosscutting concepts are similar to those in the Unifying Concepts and Processes in *NSES* and the Common Themes in *Benchmarks for Science Literacy* (*Benchmarks*). Although the previous documents call for the integration of these concepts with the content standards, the *Framework* specifically recommends, "Standards should emphasize all three dimensions articulated in the framework." (See Recommendation 4 in Chapter 12, p. 300.) This requirement was not only a challenge to the writers of the *NGSS* but also calls for a major change in instructional materials and assessments.

### Suggested Action

Participate in a review to determine if and how the Unifying Concepts and Processes in *NSES* and/or the Common Themes in *Benchmarks* are currently incorporated in your standards, curriculum, and instructional materials.

The list of crosscutting concepts in the *NGSS* will undoubtedly use the list in the *Framework,* making it possible to begin planning professional development to assist teachers in understanding and incorporating the concepts into their current teaching without waiting for the completion of the *NGSS.* The above review could serve as the impetus and needs assessment for the initiation and planning of the professional development. Exemplary instructional materials can serve as models and resources for the professional materials, but any adoption should await the release of the *NGSS.*

## Chapter 5

# Dimension 3: Disciplinary Core Ideas: Physical Sciences

## Overview

The physical sciences section has been organized under the following four core ideas and 13 component ideas.

**Core Idea PS1: Matter and Its Interactions**

- PS1.A: Structure and Properties of Matter
- PS1.B: Chemical Reactions
- PS1.C: Nuclear Processes

**Core Idea PS2: Motion and Stability: Forces and Interactions**

- PS2.A: Forces and Motion
- PS2.B: Types of Interactions
- PS2.C: Stability and Instability in Physical Systems

**Core Idea PS3: Energy**

- PS3.A: Definitions of Energy
- PS3.B: Conservation of Energy and Energy Transfer
- PS3.C: Relationship Between Energy and Forces
- PS3.D: Energy in Chemical Processes and Everyday Life

**Core Idea PS4: Waves and Their Applications in Technologies for Information Transfer**

- PS4.A: Wave Properties
- PS4.B: Electromagnetic Radiation
- PS4.C: Information Technologies and Instrumentation

The *Framework* introduces each core and component idea with an essential question that frames the main concept. Each component idea also contains grade band "endpoints" for the end of grades 2, 5, 8, and 12.

## Analysis

The *Framework* acknowledges that the content included in the first three physical science core ideas "parallel those identified in previous documents," including the *NSES* and *Benchmarks* (p. 103).

The authors introduce a fourth core idea, Waves and Their Applications in Technologies for Information Transfer, which "introduces students to the ways in which advances in the physical sciences during the 20th century underlie all sophisticated technologies today." In addition, the *Framework* acknowledges that "organizing science instruction around core disciplinary ideas tends to leave out the applications of those ideas" (p. 104). This core idea also provides an opportunity to stress the interplay between science and technology.

The endpoints, though not standards, will undoubtedly provide the disciplinary content that will form one of the three components in the performance standards called for in Recommendations 4 and 5 from Chapter 12.

## Suggested Action

Review the *Framework* endpoints for the physical sciences and compare them with the topics or outcomes in your curriculum and assessment. In each of these content areas, we suggest educators keep an eye toward the vertical alignment of the content and check to see that there are no missing core ideas at each grade band. Keep in mind that some local topics/outcomes will not appear in the *Framework* since one of the charges to the writers was to "identify a small set of core ideas in each of the major science disciplines" (p. 16). Educators can anticipate finding additional content in their local curriculum, much of which can and should be eliminated as the curriculum is adjusted to meet the *NGSS*.

The inclusion of the fourth core idea will require some additions to the curriculum of most schools when the *NGSS* are adopted by states and schools. Instructional materials for this core idea may not be readily available for some time.

The suggested action section for Chapter 8 contains suggestions for thinking about where and how engineering core ideas can be integrated in the science curriculum.

## Chapter 6

# Dimension 3: Disciplinary Core Ideas: Life Sciences

## Overview

The life sciences section has been organized under the following four core ideas and 14 component ideas.

**Core Idea LS1: From Molecules to Organisms: Structures and Processes**

- LS1.A: Structure and Function
- LS1.B: Growth and Development of Organisms
- LS1.C: Organization for Matter and Energy Flow in Organisms
- LS1.D: Information Processing

**Core Idea LS2: Ecosystems: Interactions, Energy, and Dynamics**

- LS2.A: Interdependent Relationships in Ecosystems
- LS2.B: Cycles of Matter and Energy Transfer in Ecosystems
- LS2.C: Ecosystem Dynamics, Functioning, and Resilience
- LS2.D: Social Interactions and Group Behavior

**Core Idea LS3: Heredity: Inheritance and Variation of Traits**

- LS3.A: Inheritance of Traits
- LS3.B: Variation of Traits

**Core Idea LS4: Biological Evolution: Unity and Diversity**

- LS4.A: Evidence of Common Ancestry and Diversity
- LS4.B: Natural Selection
- LS4.C: Adaptation
- LS4.D: Biodiversity and Humans

The *Framework* introduces each core and component idea with an essential question that frames the main concept. Each component idea also contains grade band endpoints for the end of grades 2, 5, 8, and 12.

## Analysis

The *Framework* states that the four core ideas "have a long history and solid foundation based on the research evidence established by many scientists working across multiple fields" (p. 141). The ideas draw on those identified in previous documents, including the *NSES* and *Benchmarks*, as well as numerous reports from the National Research Council, American Association for the Advancement of Science, National Assessment of Educational Progress, Trends in International Mathematics and Science Study, College Board, and others.

## Suggested Action

Review the *Framework* endpoints for the life sciences and compare them with the topics or outcomes in your school or district's curriculum. Keep in mind that some local topics/outcomes will not appear in the *Framework* since one of the charges to the writers was to "identify a small set of core ideas in each of the major science disciplines" (p. 16). Educators can anticipate finding additional content in their local curriculum, much of which can and should be eliminated as the curriculum is adjusted to meet the *NGSS*.

Be aware of the progression of the endpoints in each grade band. The *Framework* has been very attentive to the progression of ideas for each of the core ideas. The grade band or level may be different from your curriculum or from that of the *NSES* or *Benchmarks*.

## Chapter 7

# Dimension 3: Disciplinary Core Ideas: Earth and Space Sciences

## Overview

The Earth and space sciences section has been organized under the following three core ideas and 12 component ideas.

**Core Idea ESS1: Earth's Place in the Universe**

- ESS1.A: The Universe and Its Stars
- ESS1.B: Earth and the Solar System
- ESS1.C: The History of Planet Earth

**Core Idea ESS2: Earth's Systems**

- ESS2.A: Earth Materials and Systems
- ESS2.B: Plate Tectonics and Large-Scale System Interactions
- ESS2.C: The Roles of Water in Earth's Surface Processes
- ESS2.D: Weather and Climate
- ESS2.E: Biogeology

**Core Idea ESS3: Earth and Human Activity**

- ESS3.A: Natural Resources
- ESS3.B: Natural Hazards
- ESS3.C: Human Impacts on Earth Systems
- ESS3.D: Global Climate Change

## Analysis

The *Framework* authors drew from several recent projects to delineate the Earth and space sciences content, including *Earth Science Literacy Principles: The Big Ideas and Supporting Concepts of Earth Science* (Earth Science Literacy Initiative 2010), *Ocean Literacy: The Essential Principles of Ocean Science K–12* (NGS 2006), *Essential Principles and Fundamental Concepts for Atmospheric Science Literacy* (UCAR 2008), and *Climate Literacy: The Essential Principles of Climate Science* (U.S. Global Change Research Program 2009). The core ideas include a broader range of content than most previous standards documents, but fewer outcomes. The increased breadth is especially evident in the third core idea, Earth and Human Activity, which deals with the increased stress on the planet and its resources due to rapidly increasing population and global industrialization.

Although the core ideas of Earth and space science cover a broader range of ideas, when compared to most Earth and space science instructional materials, the number of topics (components) has been reduced significantly in most areas and the topic of human impact has been more significantly stressed. This shift will ultimately place a burden on teachers and curriculum specialists to modify their curriculum and course syllabi.

## Suggested Action

Begin the process of comparing your local curriculum to the endpoints for Earth and Space Sciences in the *Framework*. You may find that your curriculum or instructional materials have more topics and more detailed information or concepts than those outlined in the *Framework*. The opposite may be true for the third core idea, Earth and Human Activity, which describes how Earth's processes and human activity affect each other. Be aware of the progression of the endpoints in each grade band. The *Framework* has been very attentive to the progression of ideas for each of the core ideas. Local examples and illustrations of Earth science core ideas are excellent teaching resources. Begin to catalog them for use in the current curriculum or the revised curriculum, as it will help implement the *NGSS*.

## Chapter 8

# Dimension 3: Disciplinary Core Ideas: Engineering, Technology, and Applications of Science

## Overview

The engineering, technology, and applications of sciences section has been organized under the following two core ideas and five component ideas.

**Core Idea ETS1: Engineering Design**

- ETS1.A: Defining and Delimiting an Engineering Problem
- ETS1.B: Developing Possible Solutions
- ETS1.C: Optimizing the Design Solution

**Core Idea ETS2: Links Among Engineering, Technology, Science, and Society**

- ETS2.A: Interdependence of Science, Engineering, and Technology
- ETS2.B: Influence of Engineering, Technology, and Science on Society and the Natural World

## Analysis

While the intent of this chapter is to help students learn how science is used through the engineering design process, the two core ideas have different goals. The goal of the first idea is to help students develop an understanding of engineering design, while the second is to help them make connections among engineering, technology, and science. Although the *language* defining the process of engineering design may be new to science educators, the *ideas* are not new for many of them, particularly those at the elementary level and those using project activities in their teaching. For example, students designing and building a structure in an elementary science unit have followed the three procedures described in the Core Idea ETS1, possibly without the explicit understanding of the engineering design process and use of the terminology.

The early paragraphs in this chapter provide the essential, but limited, direction that learning engineering requires, combining the engineering practices outlined in Chapter 3 with the understanding of engineering design contained in Chapter 8 in the same way that science involves both knowledge and a set of practices.

The second core idea is an excellent complement to the engineering core ideas taught in the science curriculum since it brings together the interdependence of engineering, technology, science, and society. Readers familiar with the standards for Science in Personal and Societal Perspectives in the *NSES* will see some overlap with the core ideas in this section of the *Framework*.

The core ideas in this chapter and those in Chapter 3 dealing with engineering practices may prove to be a significant shift for science educators when the *NGSS* are officially published. Although many teachers and instructional materials rely on activities that are engineering in nature, the language and specific outcomes described in Core Ideas ETS1 and ETS2 are not normally included as part of the activities. A paradigm shift is called for that might be approached with the following professional development activities and curriculum development work.

## Suggested Action

Form study or discussion groups to read and discuss the nature of engineering using resources such as the National Academy of Engineering publication *Standards for K–12 Engineering Education?* (NRC 2010b). This and many other reports can be downloaded for free at *www.nap.edu.*

Study the definitions in Box 8-1, "Definitions of Technology, Engineering, and Applications of Science" (p. 202), at the end of the chapter to help gain clarity on the distinction between engineering and technology. Notice the connection between the two definitions. An excellent book on the nature of technology is *The Nature of Technology: What It Is and How It Evolves* (Arthur 2009).

Assemble a team to begin assessing how and where engineering core ideas might be integrated in the science curriculum at each grade band in your school or district. Some courses or units lend themselves to this integration better than others. What are they? Do new activities or units need to be added? Can some of the existing activities be modified or supplemented to provide outcomes in engineering? Where and how can the endpoints from the practices of engineering and the core ideas in this chapter be combined as parallel outcomes of modified or new activities?

Identify or plan professional development activities to immerse teachers in doing engineering design projects and gaining knowledge of the language and endpoints expected of their students.

# PART III:
# Realizing the Vision

## Chapter 9

# Integrating the Three Dimensions

## Overview

This chapter describes the process of integrating the three dimensions (practices, crosscutting concepts, and core ideas) in the *Next Generation Science Standards* (*NGSS*) and provides two examples for its writers, as well as for the writers of instructional materials and assessments. The preceding chapters described the dimensions separately to provide a clear understanding of each; this chapter recognizes the need and value of integrating them in standards and instruction. *A Framework for K–12 Science Education* (*Framework*) is specific about this task as indicated by the following statement (p. 218): "A major task for developers will be to create standards that integrate the three dimensions. The committee suggests that this integration should occur in the standards statements themselves and in performance expectations that link to the standards."

This expectation is based on the assumption that "students cannot fully understand scientific and engineering ideas without engaging in the practices of inquiry and the discourses by which such ideas are developed and refined. … At the same time, they cannot learn or show competence in practices except in the context of specific content" (p. 218).

Performance expectations are a necessary and essential component of the standard statements. These expectations describe how students will demonstrate an understanding and application of the core ideas. The chapter provides two illustrations in Table 9-1, "Sample Performance Expectations in the Life Sciences" (p. 220), and Table 9-2, "Sample Performance Expectations in the Physical Sciences" (p. 224), of what the performance expectation could look like for two core ideas.

Although it is not the function of the *Framework* or the *NGSS* to provide detailed descriptions of instruction, this *Framework* chapter offers a fairly extensive example—in narrative form—of what the integration of the three dimensions for a physical science core idea at each grade band (K–2, 3–5, 6–8, and 9–12) would look like. One of the unique features of this example is the inclusion of "boundary statements" that specify ideas that do *not* need to be included. The standard statements are expected to contain boundary statements.

## Analysis

Although Tables 9-1 and 9-2 are extensive examples of performance expectation for two core ideas, they are not a model for the format of the standards statements that appear in the *NGSS*. The practices and crosscutting concepts are only identified and not spelled out in performance language. The new integrated standards are a significant departure from those in the previous national standards documents, and they will have a huge impact on instruction, instructional materials, and assessments for science educators.

There are few, if any, examples or precedents for this type of standard. Such standards may very well prescribe the instruction and assessment that should be included in the curriculum and instructional materials. Performance expectations indicate the core idea, the practice that should be used or at least emphasized, and the crosscutting concepts that should be included. The performance for each of the dimensions comes close to describing how each should be assessed. The detailed instructional strategies and instructional materials will be left to the instructor, but the outcomes suggested by the practices will be determined by the standard statements and the associated performance expectations.

## Suggested Action

The development of instructional materials, their implementation, and the associated assessment from integrated standards is the second major shift (after the inclusion of engineering) that appears in the *NGSS*. We recommend the following general strategies to accommodate this shift:

- Conduct extensive reading, form study groups, and explore other professional development avenues to become deeply familiar with the scientific and engineering practices, the crosscutting concepts, and the core ideas in the *Framework*. The integration of the dimensions will be most effective if educators have a thorough and clear understanding of each dimension.
- Study Tables 9-1 and 9-2 and the narrative example of instruction from the physical sciences.
- Begin searching for instructional materials that explicitly integrate the three dimensions. Examples may begin to appear in professional literature such as NSTA journals. Examine and evaluate them carefully.
- Carefully study the content of a standard statement at your grade band. As a learning exercise, assemble a small team of colleagues and sketch out a series of lessons or a small unit to facilitate a group of students meeting the performance expectations in the standard. This exercise is only a sample of what will be required to meet the new performance expectations, but it will assist in your planning of longer-range activities and projects when the final version of the *NGSS* is published and adopted by your state or school district.

## Chapter 10

# Implementation: Curriculum Instruction, Teacher Development, and Assessment

### Overview

Most readers will recall that the *NSES* include standards for the components of teaching, professional development, assessment, educational programs, and educational systems. This chapter acknowledges the value of those standards and the fact that the charge to the *Framework* developers did not include a similar comprehensive assignment to provide standards or even recommendations. This chapter assumes the task of analyzing the overall education system and discusses "what must be in place in order for [each component] to align with the framework's vision" (p. 241). In doing so, it depends heavily on a number of recent reports from the NRC that reviewed the research related to each component in the *Framework*. These include *Knowing What Students Know* (Pellegrino, Chudowsky, and Glaser 2001), *Investigating the Influence of Standards* (Weiss et al. 2002), *Systems for State Science Assessments* (Wilson and Bertenthal 2006), *America's Lab Report* (Singer, Hilton, and Schweingruber 2006), *Taking Science to School* (Duschl, Schweingruber, and Shouse 2007), and *Preparing Teachers* (NRC 2010a).

After briefly describing the total education system and calling for coherence within it, the *Framework* addresses the components of curriculum and instruction, teacher development, and assessment.

The section on curriculum and instruction lists a variety of "aspects for curriculum designers to consider that are not addressed in the framework … that the committee considers important but decided would be better treated at the level of curriculum design than at the level of framework and standards" (p. 247). These include the historical, cultural, and ethical aspects of science and its applications, and the history of scientific and engineering ideas and the individual practitioners.

### Analysis

For many experienced science educators, this section of the *Framework* is the most important despite its limited treatment. The missing ingredient in the first release of the *NSES* and *Benchmarks* was the lack of extensive implementation at the state and local level. Both the *NSES* and the *Benchmarks* received a great deal of attention and some replication in state standards, but the standards for teaching, professional development, assessment, program, and systems did not receive equal emphasis. NSTA believes that for new standards to be implemented successfully, a significant emphasis must be placed on outreach and support for science educators.

The section in the *Framework* on instruction does not go into great depth on the topic and defers to the extensive discussion of the topic and the research behind it in *Taking Science to School* (Duschl, Schweingruber, and Shouse 2007). Teacher development and assessment sections are also light and depend on existing NRC reports previously listed in the overview section.

## Suggested Action

The call to integrate the practices, crosscutting concepts, and the core ideas will require a new and greater emphasis on incorporating change in all components of the system. The *NGSS* are what is to be implemented, not the *Framework*, but the task of implementation needs to start now, long before the *NGSS* are published and adopted in states and school districts. It is not the role of this guide to spell out the multiple steps and decisions that need to be made to implement a new set of standards, but that process needs to begin now! The starting points have been outlined in the previous sections.

For more information, visit the NSTA *NGSS* website (*www.nsta.org/ngss*), which provides a continuous flow of information about the *NGSS* throughout development and implementation.

## Chapter 11

# Equity and Diversity in Science and Engineering Education

## Overview

This chapter highlights the issues in achieving equity in education opportunities for all students, summarizes the research on the lack of equity in education in general and science education in particular, describes what should be available for all students in broad strokes, and makes a limited number of specific recommendations to the standards developers. The discussion of inequity of education achievement among specific demographic groups is reduced to two key areas: (1) the differences in the opportunity to learn due to inequities in schools and communities; and (2) the lack of inclusiveness in instruction to motivate diverse student populations. The research is clear that all students, with rare exceptions, have the capacity to learn complex subject matter when support is available over an extended period of time.

The *Framework* recommends that the *NGSS* (1) specify that rigorous learning goals (standards) are appropriate for all students and (2) make explicit the need for the instructional time, facilities, and teacher knowledge that can help all students achieve these goals.

On a more general but no less significant level, the *Framework* recommendations address the need to equalize the opportunity to learn. This means providing inclusive science instruction, making diversity visible, and providing multiple modes of expression. To make science instruction more inclusive, the *Framework* suggests several strategies: approaching science learning as a cultural accomplishment, relating youth discourses to scientific discourses, building on prior interest and identity, and leveraging students' cultural funds of knowledge.

The final recommendation in the chapter focuses on creating assessments that use multiple opportunities for students to express their understanding of the content in multiple contexts and avoiding culturally biased assessment instruments.

## Analysis

The *Framework* gives the critical issue of equity and diversity modest attention, but it provides a number of well-researched recommendations. Most of the recommendations in the chapter focus on instruction and cultural contexts of education more than the nature of standards. The limited attention to these issues in the *Framework*, due to the charge to the committee of writers, should in no way detract from its extreme importance.

> **Suggested Action**
>
> Schools should reexamine their progress with equity and diversity and reshape their efforts based on the specific recommendations provided in the *Framework*. There is no need to wait to address these issues until the *NGSS* are published; the issues of equity and diversity should be an ongoing agenda for all schools and teachers, and should be addressed aggressively and consistently.

## Chapter 12
# Guidance for Standards Developers

## Overview

This chapter opens with the recommendation from *Systems for State Science Assessments* (Wilson and Bertenthal 2006) that standards should be "clear, detailed, and complete; reasonable in scope; rigorously and scientifically correct, and based on sound models of student learning ... [and] should have a clear conceptual framework, describe performance expectations, and identify proficiency levels" (p. 298).

It then lists the following 13 specific recommendations for standard developers with a short discussion following each recommendation. (These recommendations are quoted directly from the *Framework*.)

1. Standards should set rigorous learning goals that represent a common expectation for all students (p. 298).
2. Standards should be scientifically accurate yet also clear, concise, and comprehensible to science educators (p. 299).
3. Standards should be limited in number (p. 300).
4. Standards should emphasize all three dimensions articulated in the framework—not only crosscutting concepts and disciplinary core ideas but also scientific and engineering practices (p. 300).
5. Standards should include performance expectations that integrate the scientific and engineering practices with the crosscutting concepts and disciplinary core ideas. These expectations should include criteria for identifying successful performance and require that students demonstrate an ability to use and apply knowledge (p. 301).
6. Standards should incorporate boundary statements. That is, for a given core idea at a given grade level, standards developers should include guidance not only about what needs to be taught but also about what does ***not*** need to be taught in order for students to achieve the standard (p. 301).
7. Standards should be organized as sequences that support students' learning over multiple grades. They should take into account how students' command of the practices, concepts, and core ideas becomes more sophisticated over time with appropriate instructional experiences (p. 302).
8. Whenever possible, the progressions in standards should be informed by existing research on learning and teaching. In cases in which insufficient research is available to inform a progression or in which there is a lack of consensus on the research findings, the progression should be developed on the basis of a reasoned argument about learning and teaching. The sequences described in the framework can be used as guidance (p. 303).
9. The committee recommends that the diverse needs of students and of states be met by developing grade band standards as an overarching common set for adoption by

multiple states. For those states that prefer or require grade-by-grade standards, a suggested elaboration on grade band standards could be provided as an example (p. 304).

10. If grade-by-grade standards are written based on the grade band descriptions provided in the framework, these standards should be designed to provide a coherent progression within each grade band (p. 305).
11. Any assumptions about the resources, time, and teacher expertise needed for students to achieve particular standards should be made explicit (p. 305).
12. The standards for the sciences and engineering should align coherently with those for other K–12 subjects. Alignment with the Common Core Standards in mathematics and English/language arts is especially important (p. 306).
13. In designing standards and performance expectations, issues related to diversity and equity need to be taken into account. In particular, performance expectations should provide students with multiple ways of demonstrating competence in science (p. 307).

## Analysis

Although specifically addressed to Achieve Inc., the group writing the *NGSS*, the recommendations provide a preview of what to expect in the standards document. The reader will notice that the 13 recommendations are closely aligned with the content of the first 11 chapters.

### Suggested Action

A few states and districts may be developing their own standards independent of the work being undertaken by Achieve Inc. To those few, the recommendations are germane and highly relevant. To the majority of readers, they are predictors of what to expect in the *NGSS*. In most cases, more attention should be paid to the previous sections where the issues that give rise to the recommendations are well articulated.

## Chapter 13

# Looking Toward the Future: Research and Development to Inform K–12 Science Education Standards

## Overview

Chapter 13 reminds the reader that the *Framework* is based on research and lays out the research agenda for the next near term (five to seven years) and the long term (seven years and beyond). The recommended agenda can be summarized with the following outline, which lists two major areas of research with a number of issues or questions under each.

I. Research to Inform Implementation and Future Revisions of the *Framework*
   A. Learning and Instruction
      1. What are the typical preconceptions that students hold about the practices, crosscutting concepts, and core ideas at the outset?
      2. What is the expected progression of understanding, and what are the predictable points of difficulty that must be overcome?
      3. What instructional interventions (e.g., curriculum materials, teaching practices, simulations or other technology tools, instructional activities) can move students along a path from their initial understanding to the desired outcome?
      4. What general and discipline-specific norms and instructional practices best engage and support student learning?
      5. How can students of both genders and of all cultural backgrounds, languages, and abilities become engaged in the instructional activities needed to move toward more sophisticated understanding?
      6. How can the individual student's understanding and progress be monitored? (p. 312–313)
   B. Learning Progressions
   C. Scientific and Engineering Practices
   D. Development of Curricular and Instructional Materials
   E. Assessment
   F. Supporting Teachers' Learning

II. Understanding the Impact of the *Framework* and Related Standards
   A. Curriculum and Instructional Materials
   B. Teacher and Administrator Development
   C. Assessment and Accountability
   D. Organizational Issues

## Analysis

Throughout the *Framework*, the reader is reminded that the document is based on a considerable body of solid education research, which is cited frequently. It should be pointed out that the National Research Council (NRC) does not do original research; it reviews and evaluates the research already completed by others. The NRC is a part of the National Academies, a private nonprofit institution that provides expert advice on some of the most pressing challenges facing the nation and the world through the publication of reports that have helped shape sound policies; inform public opinion; and advance the pursuit of science, engineering, and medicine. Several new documents are cited in this chapter, including *Learning and Instruction: A SERP (Strategic Education Research Partnership) Research Agenda* (Donovan and Pellegrino 2004), which influenced the agenda and research question on learning and instruction in the *Framework*. The questions in the report could lead to and shape local school district or university cooperative research activities.

### Suggested Action

Motivated readers may want to acquire and study the various research reports from the NRC that have been cited in the earlier chapters. As the standards are released and adoption and implementation begin, the question of why many of the changes or shifts from the previous documents and recommendations for classroom practices were made will be asked. The background research can be useful in making local and state decisions for curriculum and assessment and defending them in public and legislative settings.

The suggested action items in the previous chapters provide a host of ideas for science educators and others to gain a deep understanding of the *Framework* as a stand-alone document and as a guide to the use of the *NGSS*. We encourage you to pursue these and other opportunities with colleagues to better prepare for the new standards.

*Harold Pratt*, a former NSTA president, served as senior program officer at the National Research Council, where he helped develop the National Science Education Standards. He has also worked as executive director of curriculum for the Jefferson County Public Schools in Colorado and project director at BSCS. He has authored and published numerous books, chapters, and articles.

# References


American Association for the Advancement of Science (AAAS). 1993. *Benchmarks for science literacy.* New York: Oxford University Press.

Arthur, W. B. 2009. *The nature of technology: What it is and how it evolves.* New York: Free Press.

Bell, P., B. Lewenstein, A. W. Shouse, and M. A. Feder, eds. 2009. *Learning science in informal environments: People, places, and pursuits.* Washington, DC: National Academies Press.

Bransford, J. D., A. L. Brown, and R. J. Cocking, eds. 2000. *How people learn: Brain, mind, experience, and school.* Washington, DC: National Academies Press.

Donovan, M. S., and J. W. Pellegrino 2004. *Learning and instruction: A SERP (Strategic Education Research Partnership) research agenda.* Washington, DC: National Academies Press.

Duschl, R. A., H. A. Schweingruber, and A. W. Shouse, eds. 2007. *Taking science to school: Learning and teaching science in grades K–8.* Washington, DC: National Academies Press.

Earth Science Literacy Initiative. 2010. *Earth science literacy principles: The big ideas and supporting concepts of Earth science.* Arlington, VA: National Science Foundation. *www.earthscienceliteracy.org/es_literacy_6may10_.pdf*

Katehi, L., G. Pearson, and M. Feders, eds. 2009. *Engineering in K–12 education: Understanding the status and improving the prospects.* Washington, DC: National Academies Press.

Michaels, S., A. W. Shouse, and H. A. Schweingruber, eds. 2008. *Ready, set, science! Putting research to work in K–8 science classrooms.* Washington, DC: National Academies Press.

National Geographic Society (NGS). 2006. *Ocean literacy: The essential principles of ocean science K–12.* Washington, DC: NGS. *www.coexploration.org/oceanliteracy/documents/OceanLitChart.pdf*

National Research Council (NRC). 1996. *National science education standards.* Washington, DC: National Academies Press.

National Research Council (NRC). 2010a. *Preparing teachers: Building evidence for sound policy.* Washington, DC: National Academies Press.

National Research Council (NRC). 2010b. *Standards for K–12 engineering education?* Washington, DC: National Academies Press.

National Research Council (NRC). 2012. *A framework for K–12 science education: Practices, crosscutting concepts, and core ideas.* Washington, DC: National Academies Press.

Pellegrino, J. W., N. Chudowsky, and R. Glaser, eds. 2001. *Knowing what students know: The science and design of education assessment.* Washington, DC: National Academies Press.

Singer, S. R., M. L. Hilton, and H. A. Schweingruber, eds. 2006. *America's lab report: Investigations in high school science.* Washington, DC: National Academies Press.

University Corporation for Atmospheric Research (UCAR). 2008. *Essential principles and fundamental concepts for atmospheric science literacy.* Boulder, CO: UCAR. *http://eo.ucar.edu/asl/pdfs/ASLbrochureFINAL.pdf*

U.S. Global Change Research Program/Climate Change Science Program. 2009. *Climate literacy: The essential principles of climate science.* Washington, DC: U.S. Global Change Research Program/Climate Change Science Program. *www.climatescience.gov/Library/Literacy/default.php*

Weiss I. R., M. S. Knapp, K. S. Hollweg, and G. Burrill, eds. 2002. *Investigating the influence of standards: A framework for research in mathematics, science, and technology education.* Washington, DC: National Academies Press.

Wilson, M. R., and M. W. Bertenthal, eds. 2006. *Systems for state science assessments.* Washington, DC: National Academies Press.

# PART IV:
## Understanding *A Framework for K–12 Science Education:* Top Science Educators Offer Insight

# Scientific and Engineering Practices in K–12 Classrooms

***By Rodger W. Bybee***

This morning I watched *Sesame Street*. During the show, characters "acted like engineers" and designed a boat so a rock could float. In another segment, children asked questions and made predictions about the best design for a simple car. They then built a model car and completed an investigation to determine which design worked best when the cars went down inclined planes. Children also learned that a wider base provided stability for a tower. And, among other segments, the children counted from 1 to 12 and explored the different combinations of numbers that equaled 12. Bert and Ernie had to move a rock and ended up "inventing" a wheel. These segments exemplify the science, technology, engineering, and mathematics (STEM) theme that *Sesame Street* is introducing in the show's 42nd season.

What, you ask, does this have to do with science and engineering practices in K–12 classrooms? The producers of *Sesame Street* decided that STEM practices were important enough that they are using them as substantive themes for the season, if not longer. Children watching *Sesame Street* will have been introduced to practices such as asking questions and defining problems; developing and using models; planning and carrying out investigations; analyzing and interpreting data; using mathematics; constructing explanations and designing solutions; engaging in arguments using evidence; and obtaining, evaluating, and communicating information. True, these are sophisticated statements of practices, but many students will be introduced to them when they enter elementary classrooms.

Here, I present the science and engineering practices from the recently released *A Framework for K–12 Science Education: Practices, Crosscutting Concepts, and Core Ideas* (*Framework*; NRC 2012). I recognize the changes implied by the new framework, and eventually a new generation of science education standards will present new perspectives for the science education community. I am especially sensitive to the challenges for those students in teacher preparation programs and classroom teachers of science at all levels. Questions such as "Why practices and why not inquiry?" and "Why science *and* engineering?" are reasonable, and I will discuss them later. But to provide background and context, I first discuss the practices.

## Understanding and applying the science and engineering practices

This section further elaborates on the practices and briefly describes what students are to know and be able to do and how they might be taught. Figures 1 through 8 are adapted from the National Research Council (NRC) *Framework*, with changes for clarity and balance. I have maintained the substantive content.

Even before elementary school, children ask questions of each other and of adults about things around them, including the natural and designed world. If students develop the practices of science and engineering, they can ask better questions and improve how they define problems. Students should, for example, learn how to ask questions of each other, to recognize the difference between questions and problems, and to evaluate scientific questions and engineering problems from other types of questions. In upper grades, the practices of asking scientific questions and defining engineering problems advance in subtle ways such as the form and function of data used in answering questions and the criteria and constraints applied to solving problems.

In the lower grades, the idea of scientific and engineering models can be introduced using pictures, diagrams, drawings, and simple physical models such as airplanes or cars. In upper grades, simulations and more sophisticated conceptual, mathematical, and computational models may be used to conduct investigations, explore changes in system components, and generate data that can be used in formulating scientific explanations or in proposing technological solutions.

Planning and carrying out investigations should be standard experiences in K–12 classrooms. Across the grades students develop deeper and richer understandings and abilities as they conduct different types of investigations, use different technologies to collect data, give greater attention to the types of variables, and clarify the scientific and/or engineering contexts for investigations.

**Figure 1. Asking questions and defining problems**

| | |
|---|---|
| **Science** begins with a question about a phenomenon such as "Why is the sky blue?" or "What causes cancer?" A basic practice of the scientist is the ability to formulate empirically answerable questions about phenomena to establish what is already known, and to determine what questions have yet to be satisfactorily answered. | **Engineering** begins with a problem that needs to be solved, such as "How can we reduce the nation's dependence on fossil fuels?" or "What can be done to reduce a particular disease?" or "How can we improve the fuel efficiency of automobiles?" A basic practice of engineers is to ask questions to clarify the problem, determine criteria for a successful solution, and identify constraints. |

**Figure 2. Developing and using models**

| | |
|---|---|
| **Science** often involves the construction and use of models and simulations to help develop explanations about natural phenomena. Models make it possible to go beyond observables and simulate a world not yet seen. Models enable predictions of the form "if...then...therefore" to be made in order to test hypothetical explanations. | **Engineering** makes use of models and simulations to analyze extant systems to identify flaws that might occur, or to test possible solutions to a new problem. Engineers design and use models of various sorts to test proposed systems and to recognize the strengths and limitations of their designs. |

**Figure 3. Planning and carrying out investigations**

| | |
|---|---|
| **Scientific investigations** may be conducted in the field or in the laboratory. A major practice of scientists is planning and carrying out systematic investigations that require clarifying what counts as data and in experiments identifying variables. | **Engineering investigations** are conducted to gain data essential for specifying criteria or parameters and to test proposed designs. Like scientists, engineers must identify relevant variables, decide how they will be measured, and collect data for analysis. Their investigations help them to identify the effectiveness, efficiency, and durability of designs under different conditions. |

Both science and engineering involve the analysis and interpretation of data. In lower grades, students simply record and share observations though drawings, writing, whole numbers, and oral reports. In middle and high school, students report relationships and patterns in data, distinguish between correlation and causation, and compare and contrast independent sets of data for consistency and confirmation of an explanation or solution.

The overlap of these practices with the next practices, using mathematical and computational thinking, is significant. Although both of these sets of practices can be completed with simulated data, it is beneficial for students to actually experience the practices of collecting, analyzing, and interpreting data and in the process apply mathematical and computational thinking.

In the early grades, students can learn to use appropriate instruments (e.g., rulers and thermometers) and their units in measurements and in quantitative results to compare proposed solutions to an engineering problem. In upper grades, students can use computers to analyze data sets and express the significance of data using statistics.

Students can learn to use computers to record measurements, summarize and display data, and calculate relationships. As students progress to higher grades, their experiences in science classes should enhance what they learn in math class.

The aim for students at all grade levels is to learn how to use evidence to formulate a logically coherent explanation of phenomena and to support a proposed solution for an engineering problem. The construction of an explanation or solution should incorporate current

**Figure 4. Analyzing and interpreting data**

| | |
|---|---|
| **Scientific investigations** produce data that must be analyzed in order to derive meaning. Because data usually do not speak for themselves, scientists use a range of tools—including tabulation, graphical interpretation, visualization, and statistical analysis—to identify the significant features and patterns in the data. Sources of error are identified and the degree of certainty calculated. Modern technology makes the collection of large data sets much easier, providing secondary sources for analysis. | **Engineering investigations** include analysis of data collected in the tests of designs. This allows comparison of different solutions and determines how well each meets specific design criteria—that is, which design best solves the problem within given constraints. Like scientists, the engineers require a range of tools to identify the major patterns and interpret the results. Advances in science make analysis of proposed solutions more efficient and effective. |

**Figure 5. Using mathematics and computational thinking**

| | |
|---|---|
| In **science,** mathematics and computation are fundamental tools for representing physical variables and their relationships. They are used for a range of tasks such as constructing simulations; statistically analyzing data; and recognizing, expressing, and applying quantitative relationships. Mathematical and computational approaches enable prediction of the behavior of physical systems along with the testing of such predictions. Moreover, statistical techniques are also invaluable for identifying significant patterns and establishing correlational relationships. | In **engineering,** mathematical and computational representations of established relationships and principles are an integral part of the design process. For example, structural engineers create mathematical-based analysis of designs to calculate whether they can stand up to expected stresses of use and if they can be completed within acceptable budgets. Moreover, simulations provide an effective test bed for the development of designs as proposed solutions to problems and their improvement, if required. |

scientific knowledge and often include a model. These practices along with those in Figure 1 differentiate science from engineering.

In elementary grades, students might listen to two different explanations for an observation and decide which is better supported with evidence. Students might listen to other students' proposed solutions and ask for the evidence supporting the proposal. In upper grades, students should learn to identify claims; differentiate between data and evidence; and use logical reasoning in oral, written, and graphic presentations.

**Figure 6. Constructing explanations and designing solutions**

| | |
|---|---|
| The goal of **science** is the construction of theories that provide explanatory accounts of the material world. A theory becomes accepted when it has multiple independent lines of empirical evidence, greater explanatory power, a breadth of phenomena it accounts for, and explanatory coherence and parsimony. | The goal of **engineering** design is a systematic solution to problems that is based on scientific knowledge and models of the material world. Each proposed solution results from a process of balancing competing criteria of desired functions, technical feasibility, cost, safety, aesthetics, and compliance with legal requirements. Usually there is no one best solution, but rather a range of solutions. The optimal choice depends on how well the proposed solution meets criteria and constraints. |

**Figure 7. Engaging in argument from evidence**

| | |
|---|---|
| In **science**, reasoning and argument are essential for clarifying strengths and weaknesses of a line of evidence and for identifying the best explanation for a natural phenomenon. Scientists must defend their explanations, formulate evidence based on a solid foundation of data, examine their understanding in light of the evidence and comments by others, and collaborate with peers in searching for the best explanation for the phenomena being investigated. | In **engineering**, reasoning and argument are essential for finding the best solution to a problem. Engineers collaborate with their peers throughout the design process. With a critical stage being the selection of the most promising solution among a field of competing ideas. Engineers use systematic methods to compare alternatives, formulate evidence based on test data, make arguments to defend their conclusions, critically evaluate the ideas of others, and revise their designs in order to identify the best solution. |

**Figure 8. Obtaining, evaluating, and communicating information**

| | |
|---|---|
| **Science** cannot advance if scientists are unable to communicate their findings clearly and persuasively or learn about the findings of others. A major practice of science is thus to communicate ideas and the results of inquiry—orally; in writing; with the use of tables, diagrams, graphs, and equations; and by engaging in extended discussions with peers. Science requires the ability to derive meaning from scientific texts such as papers, the internet, symposia, or lectures to evaluate the scientific validity of the information thus acquired and to integrate that information into proposed explanations. | **Engineering** cannot produce new or improved technologies if the advantages of their designs are not communicated clearly and persuasively. Engineers need to be able to express their ideas orally and in writing; with the use of tables, graphs, drawings, or models; and by engaging in extended discussions with peers. Moreover, as with scientists, they need to be able to derive meaning from colleagues' texts, evaluate information, and apply it usefully. |

In elementary grades, these practices entail sharing scientific and technological information; mastering oral and written presentations; and appropriately using models, drawings, and numbers. As students progress, the practices become more complex and might include preparing reports of investigations; communicating using multiple formats; constructing arguments; and incorporating multiple lines of evidence, different models, and evaluative analysis.

With this introduction and overview of science and engineering practices, I turn to some of the questions engaged by a shift in teaching strategies and learning outcomes. Although science teachers have many questions, the next sections discuss two questions that seem prominent: "Why *practices*?" and "Why *engineering*?"

## Why *practices?*

Science teachers have asked, "Why use the term *practices*? Why not continue using *inquiry*?" These are reasonable questions. A brief history will provide the context for an answer.

One major innovation in the 1960s reform movement was the introduction of the *processes* of science as a replacement for the *methods* of science. The processes of science shifted the emphasis from students' memorizing five steps in the scientific method to learning specific and fundamental processes such as observing, clarifying, measuring, inferring, and predicting. To complement this new emphasis, the new reformed instructional materials incorporated activities, laboratories, and investigations that gave students opportunities to learn the processes of science while developing an understanding of the conceptual structure of science disciplines.

During the period 1960–1990, interest and support grew for *scientific inquiry* as an approach to science teaching that emphasized learning science concepts and using the skills and abilities of inquiry to learn those concepts.

This change toward scientific inquiry was expressed by leaders such as Joseph Schwab and Paul Brandwein and publications such as *Science for All Americans* (Rutherford and Ahlgren 1989). In the 1990s, scientific inquiry was fundamental to the *Benchmarks for Science Literacy* (*Benchmarks*; AAAS 1993) and the *National Science Education Standards* (*NSES*; NRC 1996). Along with *Inquiry and the National Science Education Standards* (NRC 2000), these two publications had a significant influence on state standards and the place of inquiry in school science programs. It is important that scientific inquiry expanded and improved the earlier processes of science and provided richer understanding of science, a set of cognitive abilities for students, and more effective teaching strategies. One should note that the reforms toward the *processes of science* and *scientific inquiry* did result in greater emphasis on the use of activities and investigations as teaching strategies to learn science concepts. However, scientific inquiry has not been implemented as widely as expected.

During the 15 years since the release of the standards, researchers have advanced our knowledge about how students learn science (Bybee 2002; Donovan and Bransford 2005) and the way science functions. Advances in these and other areas have been synthesized in *Taking Science to School* (Duschl, Schweingruber, and Shouse 2007) and *Ready, Set, Science!* (Michaels, Shouse, and Schweingruber 2008). These two publications had a significant influence on the *Framework*.

*Taking Science to School* describes four proficiencies that link the content and practices of science. "Students who are proficient in science," the authors write,

- *know, use, and interpret scientific explanations of the natural world;*
- *generate and evaluate scientific evidence and explanations;*
- *understand the nature and development of scientific knowledge; and*
- *participate productively in scientific practices and discourse.* (Duschl, Schweingruber, and Shouse 2007, p. 2)

The following quote from *Ready, Set, Science!* builds on these proficiencies and presents an answer to the question, "Why practices?"

> *Throughout this book, we talk about "scientific practices" and refer to the kind of teaching that integrates the four strands as "science as practice." Why not use the term "inquiry" instead? Science practice involves doing something and learning something in such a way that the doing and learning cannot really be separated. Thus, "practice" . . . encompasses several of the different dictionary definitions of the term. It refers to doing something repeatedly in order to become proficient (as in practicing the trumpet). It refers to learning something so thoroughly that it becomes second nature (as in practicing thrift). And it refers to using one's knowledge to meet an objective (as in practicing law or practicing teaching).* (Michaels, Shouse, and Schweingruber 2008, p. 34)

Scientific inquiry is one form of scientific practice. So, the perspective presented in the *Framework* is not one of replacing inquiry; rather, it is one of expanding and enriching the teaching and learning of science. Notice the emphasis on teaching strategies aligned with science practices. When students engage in scientific practices, activities become the basis for learning about experiments, data and evidence, social discourse, models and tools, and mathematics and for developing the ability to evaluate knowledge claims, conduct empirical investigations, and develop explanations.

## Why *engineering?*

Again, a brief history establishes a context for the inclusion of engineering practices. In the 1960s, technology and engineering were marginalized in the U.S. science curriculum (Rudolph 2002). This said, the era of curriculum reform in the United States did produce one program, *The Man Made World*, developed by the Engineering Concepts Curriculum Project (1971). However, technology was included in other countries (Black and Atkin 1996; Atkin and Black 2003). Publication of *Science for All Americans* (Rutherford and Ahlgren 1989) included chapters on "the nature of technology" and "the Designed World." This reintroduction of technology and engineering was further advanced by their inclusion in the *Benchmarks* and *NSES*. Technology gained further support with the publication of the *Standards for Technological Literacy* (ITEA 2000).

In the early 21st century, the acronym STEM has emerged as a description of many and diverse educational initiatives. The *T* and *E* in STEM represent *technology* and *engineering.*

As the reader no doubt recognized in the eight figures, the practices of science and engineering overlap in many ways. With the exception of their goals—science proposes questions about the natural world and proposes answers in the form of evidence-based explanations, and engineering identifies problems of human needs and aspirations and proposes solutions in the form of new products and processes—science and engineering practices are parallel and complementary.

So, there is a need for science teachers and those in teacher education programs to recognize the similarities and differences between science and technology as disciplines and subsequently the practices that characterize the disciplines.

At elementary levels, there is good news. Many activities that are already in the curriculum are based on engineering problems. Building bridges, dropping eggs, and (as we saw in the opening on *Sesame Street*) designing model cars are all examples of engineering in elementary school programs. Unfortunately, these engineering problems and subsequent practices are often referred to erroneously as science. With a clarification of terms and a continuation of the activities, elementary teachers can introduce science and engineering practices without significant additions to the curriculum. And, as value added, the engineering problems are highly motivating for the students at all grade levels.

At the middle and high school levels, science teachers can begin with the technologies already used—microscopes, telescopes, and computers—as examples of the relationship between science and technology. In addition, there are examples clearly embedded in the practices of science and engineering. Here, I would also add the value of the history of science to show the role of technology and engineering and their contributions to the advance of scientific knowledge. An excellent contemporary example of the advance of science that is due to technology and engineering is the Hubble Space Telescope and its potential successor, the James Webb Space Telescope.

## Complementing goals

This article explores one aspect of the new NRC *Framework*—science and engineering practices—in greater depth. Although the NRC report is a framework and not standards, it is prudent for those in the science and technology education community to begin preparing for the new standards.

Because science and engineering practices are basic to science education and the change from inquiry to practices is central, this innovation for the new standards will likely be one of the most significant challenges for the successful implementation of science education standards. The brief discussion that follows is based on the prior description of science and engineering practices in Figures 1 through 8.

The relationship between science and engineering practices is one of complementarity. Given the inclusion of engineering in the science standards and an understanding of the difference in aims, the practices complement one another and should be mutually reinforcing in curricula and instruction.

The shift to practices emerges from research on how students learn and advances our understanding of how science progresses. The new emphasis on practices includes scientific inquiry and goes beyond what science teachers have realized based on the 1990s standards. Indeed, as I have noted, there is overlap with the 1996 standards, for example.

The new emphasis on practices reinforces the need for school science programs to actively involve students through investigations and, in the 21st century, digitally based programs and activities. Hands-on and laboratory work should still contribute to the realization of practices in science classrooms. As we saw in the earlier quote from *Ready, Set, Science!,* there is a reasonable assumption that across the K–12 continuum the abilities and understandings of science and engineering practices will progressively get deeper and broader.

Science and engineering practices should be thought of as both learning outcomes and instructional strategies. They represent both educational ends and instructional means. First, students should develop the abilities described in the practices, and they should understand how science knowledge and engineering products develop as a result of the practices. Second, as instructional strategies, the practices provide a means to the learning outcomes just described and other valued outcomes such as students' understanding of the core ideas and crosscutting concepts expressed in the *Framework*. In brief, the practices represent one aspect of what students are to know, what they are able to do, and how they should be taught. Granted, this is a large order, but from the perspective of K–12, teachers will have 13 years to facilitate students' attaining the goals.

To conclude, watching the children and characters on *Sesame Street* gave me confidence that the new challenges are achievable and that K–12 science education will have a generation of boys and girls ready to engage in and learn from science and engineering practices. Preparing for the next generation of science education standards will help science teachers attain the higher aspiration of this and future generations.

*Rodger W. Bybee* is executive director emeritus of Biological Sciences Curriculum Study (BSCS).

## References

American Association for the Advancement of Science (AAAS). 1993. *Benchmarks for Science Literacy.* Washington, DC: AAAS.

Atkin, J. M., and P. Black. 2003. *Inside science education reform: A history of curricular and policy change.* New York: Teachers College Press, Columbia University.

Black, P., and J. M. Atkin, eds. 1996. *Changing the subject: Innovations in science, mathematics and technology education.* London: Routledge.

Bybee, R. W., ed. 2002. *Learning science and the science of learning.* Arlington, VA: NSTA Press.

Donovan, S., and J. Bransford, eds. 2005. *How students learn: Science in the classroom.* Washington, DC: National Academies Press.

Duschl, R., H. Schweingruber, and A. Shouse, eds. 2007. *Taking science to school: Learning and teaching science in grades K–8.* Washington, DC: National Academies Press.

Engineering Concepts Curriculum Project. 1971. *The man made world.* New York: McGraw Hill.

International Technology Education Association (ITEA). 2000. *Standards for technological literacy: Content for the study of technology.* Reston, VA: ITEA.

Michaels, S., A. Shouse, and H. Schweingruber. 2008. *Ready, set, science!: Putting research to work in K–8 science classrooms.* Washington, DC: National Academies Press.

National Research Council (NRC). 1996. *National science education standards.* Washington, DC: National Academies Press.

National Research Council (NRC). 2000. *Inquiry and the national science education standards.* Washington, DC: National Academies Press.

National Research Council (NRC). 2012. *A framework for K–12 science education: Practices, crosscutting concepts, and core ideas.* Washington, DC: National Academies Press.

Rudolph, J. L. 2002. *Scientists in the classroom: The Cold War reconstruction of American science education.* New York: Palgrave Macmillan.

Rutherford, F. J., and A. Ahlgren. 1989. *Science for all Americans.* New York: Oxford University Press.

# Core Ideas of Engineering and Technology

*By Cary Sneider*

Rodger Bybee's chapter, "Scientific and Engineering Practices in K–12 Classrooms," provided an overview of Chapter 3 in *A Framework for K–12 Science Education: Practices, Crosscutting Concepts, and Core Ideas* (*Framework*; NRC 2012). Chapter 3 describes the practices of science and engineering that students are expected to develop during 13 years of schooling and emphasizes the similarities between science and engineering.

This essay addresses Chapter 8 of the *Framework,* which presents core ideas in technology and engineering at the same level as core ideas in the traditional science fields, such as Newton's laws of motion and the theory of biological evolution. Although prior standards documents included references to engineering and technology, they tended to be separate from the "core content" of science, so they were often overlooked.

Giving equal status to engineering and technology raises a number of important issues for curriculum developers and teachers, a few of which I will discuss here:

- How does the *Framework* define *science, engineering,* and *technology?*
- What are the core ideas in Chapter 8?
- Why is there increased emphasis on engineering and technology?
- Is it redundant to have engineering practices *and* core ideas?
- Do we need to have special courses to teach these core ideas?
- Will teachers need special training?
- What will it look like in the classroom?

## How does the *Framework* define *science, engineering,* and *technology?*

The meanings of these terms are summarized in the first chapter of the *Framework* as follows:

> *In the K–12 context, science is generally taken to mean the traditional natural sciences: physics, chemistry, biology, and (more recently) Earth, space, and environmental sciences. . . . We use the term engineering in a very broad sense to mean any engagement in a systematic practice of design to achieve solutions to particular human problems. Likewise, we broadly use the term technology to include all types of human-made systems and processes—not in the limited sense often used in schools that equates technology with modern computational and communications devices. Technologies result when engineers apply their understanding of the natural world and of human behavior to design ways to satisfy human needs and wants.* (NRC 2012, pp. 11–12)

Notice that engineering is *not* defined as applied science. Although the practices of engineering have much in common with the practices of science, engineering is a distinct field and has certain core ideas that are different from those of science. Given the need to limit the number of standards so that the task for teachers and students is manageable, just two core ideas are proposed in Chapter 8. The first concerns ideas about engineering design that were not addressed in Chapter 3, and the second concerns the links among engineering, technology, science, and society.

## What are the core ideas in Chapter 8?

As with core ideas in the major science disciplines, the two core ideas related to engineering and technology are first stated broadly, followed by grade band endpoints to specify what additional aspects of the core idea students are expected to learn at each succeeding level. Following are brief excerpts from the rich descriptions in the *Framework:*

***Core Idea 1: Engineering Design***

*From a teaching and learning point of view, it is the iterative cycle of design that offers the greatest potential for applying science knowledge in the classroom and engaging in engineering practices. The components of this core idea include understanding how engineering problems are defined and delimited, how models can be used to develop and refine possible solutions to a design problem, and what methods can be employed to optimize a design.* (NRC 2012, pp. 201–202)

- By the end of second grade, students are expected to understand that engineering problems may have more than one solution and that some solutions are better than others.
- By the end of fifth grade, students are expected to be able to specify problems in terms of criteria for success and constraints, or limits, to understand that when solving a problem it is important to generate several different design solutions by taking relevant science knowledge into account and to improve designs through testing and modification. In some cases it is advisable to push tests to the point of failure to identify weak points.
- By the end of middle school, students should be able to recognize when it makes sense to break complex problems into manageable parts; to systematically evaluate different designs, combining the best features of each; to conduct a series of tests to refine and optimize a design solution; and to conduct simulations to test if–then scenarios.
- By the time they graduate from high school, students should be able to do all of the above and, in addition, formulate a problem with quantitative specifications; apply knowledge of both mathematics and science to develop and evaluate possible solutions; test designs using mathematical, computational, and physical models; and have opportunities to analyze the way technologies evolve through a research and development cycle.

Core Idea 2 (Links Among Engineering, Technology, Science, and Society) has two components that are more distinct than the three components of engineering design, so they are listed separately.

***Core Idea 2A: Interdependence of Science, Engineering, and Technology***

*The fields of science and engineering are mutually supportive. New technologies expand the reach of science, allowing the study of realms previously inaccessible to investigation; scientists depend on the work of engineers to produce the instruments and computational tools they need to conduct research. Engineers in turn depend on the work of scientists to understand how different technologies work so they can be improved; scientific discoveries are exploited to create new technologies in the first place. Scientists and engineers often work together in teams, especially in new fields, such as nanotechnology or synthetic biology that blur the lines between science and engineering.* (NRC 2012, p. 203)

- By the end of second grade, students should know that engineers design a great many different types of tools that scientists use to make observations and measurements. Engineers also make observations and measurements to refine solutions to problems.
- By the end of fifth grade, students learn more about the role played by engineers in designing a wide variety of instruments used by scientists (e.g., balances, thermometers, graduated cylinders, telescopes, and microscopes). They also learn that scientific discoveries have led to the development of new and improved technologies.
- By the end of eighth grade, students learn that engineering advances have led to the establishment of new fields of science and entire industries. They also learn that the need to produce new and improved technologies (such as sources of energy that do not rely on fossil fuels and vaccines to prevent disease) have led to advances in science.
- By the time they graduate from high school, students should be aware of how scientists and engineers who have expertise in a number of different fields work together to solve problems to meet society's needs.

***Core Idea 2B: Influence of Engineering, Technology, and Science on Society and the Natural World***

*The applications of science knowledge and practices to engineering, as well as to such areas as medicine and agriculture, have contributed to the technologies and the systems that support them that serve people today. . . . In turn, society influences science and engineering. Societal decisions, which may be shaped by a variety of economic, political, and cultural factors, establish goals and priorities for technologies' improvement or replacement. Such decisions also set limits—in controlling the extraction of raw materials, for example, or in setting allowable emissions of pollution from mining, farming, and industry.* (NRC 2012, p. 202)

- By the end of second grade, students recognize that their lives depend on various technologies and that life would be very different if those technologies were to disappear. They also understand that all products are made from natural materials and that creating and using technologies have impacts on the environment.
- By the end of fifth grade, students realize that as people's needs and wants change so do their demands for new and improved technologies that drive the work of

engineers. And when those new technologies are developed, they may bring about changes in the ways that people live and interact with each other.

- By the end of eighth grade, students are familiar with cases in which the development of new and improved technologies has had both positive and negative impacts on people and the environment. They understand that the development of new technologies is driven by individual and societal needs as well as by scientific discoveries and that available technologies differ from place to place and over time because of such factors as culture, climate, natural resources, and economic conditions.
- By the time they graduate from high school, students are aware of the major technological systems that support modern civilization; how engineers continually modify these systems to increase benefits while decreasing risks; and how adoption of new technologies depends on such factors as market forces, societal demands, and government support or regulation. By the end of 12th grade, students should be able to analyze costs and benefits so as to inform decisions about the development and use of new technologies.

## Why is there increased emphasis on engineering and technology?

The commitment to engineering and technology in the *Framework* is extensive, as references to these terms are found throughout the document. A rationale for this increased emphasis is stated in different ways at a number of places in the *Framework*. One reason is inspirational, as described in the following paragraph:

> *We anticipate that the insights gained and interests provoked from studying and engaging in the practices of science and engineering during their K–12 schooling should help students see how science and engineering are instrumental in addressing major challenges that confront society today, such as generating sufficient energy, preventing and treating diseases, maintaining supplies of clean water and food, and solving the problems of global environmental change. In addition, although not all students will choose to pursue careers in science, engineering, or technology, we hope that a science education based on the Framework will motivate and inspire a greater number of people—and a better representation of the broad diversity of the American population—to follow these paths than is the case today.* (NRC 2012, p. 9)

A second reason is practical. The value of developing useful knowledge and skills is summed up in the following:

> *First, the committee thinks it is important for students to explore the practical use of science, given that a singular focus on the core ideas of the disciplines would tend to shortchange the importance of applications. Second, at least at the K–8 level, these topics typically do not appear elsewhere in the curriculum and thus are neglected if not included in science instruction. Finally, engineering and technology provide a context in which students can test their*

*own developing scientific knowledge and apply it to practical problems; doing so enhances their understanding of science—and, for many, their interest in science—as they recognize the interplay among science, engineering, and technology. We are convinced that engagement in the practices of engineering design is as much a part of learning science as engagement in the practices of science.* (NRC 2012, p. 12)

## Is it redundant to have engineering practices *and* core ideas?

This is an excellent question, especially because there is no corresponding chapter about core ideas of scientific inquiry. However, a close reading of the *Framework* will reveal that although there is some overlap between Chapter 3 and Chapter 8, very little of the content is redundant. Chapter 3 treats engineering design as a set of practices that are similar to scientific inquiry so that students may develop these abilities in the context of asking and answering questions about the world as well as systematically solving problems. Chapter 8 expands on engineering design in ways not mentioned in Chapter 3, addressing such issues as systematically evaluating potential solutions, testing to failure, and the process of optimization.

Also, a major focus of Chapter 8 concerns the interrelationships among science, engineering, technology, society, and the environment, which are essential for all students and are not addressed anywhere else in the document. An important message of this set of core ideas is that it is important for everyone not only to know how to design technological solutions to problems, but also to think broadly about the potential impacts of new and improved technologies and to recognize the role and responsibility that all citizens have to guide the work of scientists and engineers by the decisions they make as workers, consumers, and citizens.

## Do we need to have special courses to teach these core ideas?

The *Framework* provides a broad description of the content and sequence of learning expected of all students but does not provide grade-by-grade standards or specify courses at the high school level. There are many ways that these ideas can be combined and presented using a wide variety of media and learning activities. Schools are not asked to offer courses entitled "Engineering" or "Technology" any more than they are asked to offer courses with the title "Scientific Inquiry," although they may certainly do so. And although the *Next Generation Science Standards* (Achieve Inc. 2013), which is based on the *Framework*, specifies learning standards at a finer level of detail, it does not recommend specific courses.

## Will teachers need special training?

Many of the ideas about engineering and technology in the *Framework* will be familiar to today's science teachers. Many science curriculum materials include practical applications of science concepts and provide design challenges alongside science inquiry activities. Subjects such as circuit electricity and simple machines, which fall squarely in the realm of technology, have traditionally been a part of the science curriculum.

However, there will be subtle but important differences that teachers will need to become aware of. For example, design challenges are commonly presented without specific instruction in engineering design principles. Although students may have a good time and come up with creative solutions, without specific guidance they are not likely to learn about the value of defining problems in terms of criteria and constraints, how to use the problem definition to systematically evaluate alternative solutions, how to construct and test models, how to use failure analysis, or how to prioritize constraints and use trade-offs to optimize a design. Consequently, it will take some time for curriculum developers and teachers to learn about the new features of the *Framework* and incorporate these ideas into their practices. Undoubtedly the process will be greatly facilitated by inservice professional development as well as modifications of preservice preparation programs for new teachers.

## What will it look like in the classroom?

There are innumerable examples in existing curricula that illustrate engineering and technology instruction at all grade levels, many in conjunction with lessons in the natural sciences. An extensive database of materials with expert teacher reviews is available via the web at the National Center for Technological Literacy (2011), hosted by the Museum of Science in Boston. The free website, called the Technology & Engineering Curriculum Review, provides a search engine that lets teachers search by grade level, topic, or science standards to find relevant materials.

Because selecting any one of the existing materials as an example would be unfair to all the others, I've chosen to close this article with an invented example, to illustrate how the teaching of science might be enriched with an engineering activity.

Imagine a physical science class in which students are being introduced to Newton's third law, which states that every action has an equal and opposite reaction. The teacher blows up a balloon then lets it go. The balloon flies wildly around the room as air escapes out of the back end. The students are challenged to use Newton's third law to explain why the balloon flew around the room. If the students understand the basic concept, the teacher might go on to have students solve numerical problems involving Newton's third law or introduce a different topic.

Expanding on the lesson with an engineering design challenge is one way to introduce the relationship between science and engineering and to engage students in applying other concepts that they learned earlier in the year. Following the previous lesson, imagine that the teacher now asks the students to modify the balloon so that it flies more like a proper rocket—on a straight, predictable course, with as much speed and distance as possible—applying other appropriate science concepts learned previously.

Do they need to use the balloon the teacher gave them, or could they use one made from thicker rubber so they could increase the air pressure inside the balloon? Could they attach a straw and string to guide its path, or would the rocket need to fly freely? Teams would be urged to generate a number of design ideas and to evaluate them on the basis of the criteria and constraints of the problem. They would be urged to consider trade-offs as part of their planning effort; to test their designs, carefully controlling variables to determine which design works best; and to communicate the solution along with the test results that provide evidence in support of the optimal design.

Adding an engineering design challenge like the one previously described will add time to the lesson. That is not necessarily a bad thing if the science concept being applied is important to teach and challenging for students to understand without concrete examples. There are also many other approaches to introducing engineering and technology into science lessons, such as conducting research on the internet or discussing relevant current events that require less time and may focus on more important issues. And, of course, not all science ideas lend themselves easily to engineering and technology connections.

No matter how carefully new curriculum materials are designed, however, some additional time will be needed for students to apply what they are learning to the real world. Today's science curriculum is so packed that it is difficult to imagine how to add yet another set of ideas on top of what we have now. Consequently, our greatest challenge as a profession will not be whether or how to integrate engineering and technology into the curriculum, because most science educators have long considered these ideas to be an essential part of what they already do. Instead, the challenge will be how to make the difficult choices about what can safely be left out of the curriculum, so that we can do a better job of teaching core ideas and helping our students understand why they are important and how to apply them to real problems.

*Cary Sneider* is an associate research professor at Portland State University, Portland, Oregon. He served as the Design Team's Lead for Engineering and Technology during the development of the *Framework*.

## References

Achieve Inc. 2013. *Next generation science standards. www.nextgenscience.org/next-generation-science-standards*

National Center for Technological Literacy. 2011. Technology & Engineering Resources. Boston: Museum of Science. *www.mos.org/TEC*

National Research Council (NRC). 2012. *A framework for K–12 science education: Practices, crosscutting concepts, and core ideas.* Washington, DC: National Academies Press.

# The Second Dimension—Crosscutting Concepts

*By Richard A. Duschl*

For the last half century educators have struggled with the question, "What do we want students to know and what do they need to do to know it?" An alternative perspective for planning and framing science instruction asks "What do we want students to do and what do they need to know to do it?" The recently published National Research Council (NRC) report *A Framework for K–12 Science Education: Practices, Crosscutting Concepts, and Core Ideas* (*Framework*; NRC 2012) offers a thoughtful research-based agenda that helps guide us in making the shift to a doing-led agenda in K–12 science education. Grounded in the recommendations and conclusions from the NRC research synthesis report, *Taking Science to School* (NRC 2007), which I chaired, the *Framework* proposes that:

1. K–12 science education be coordinated around three intertwining dimensions: practices, crosscutting concepts, and core ideas; and
2. curricula, instruction, and assessments be aligned and then coordinated across grade band learning progressions.

In "Scientific and Engineering Practices in K–12 Classrooms," Rodger Bybee focused on scientific and engineering practices, dimension one of the *Framework*. Here the focus is on the *Framework*'s crosscutting concepts—dimension two. The *Framework* makes very clear that science learning needs to be coordinated around generative conceptual ideas and scientific practices. I begin with the seven crosscutting concepts, highlighting features within each that reveal the components of progressions. A big challenge for teachers is thinking about planning lessons and units across grade bands as student learning progresses within a grade and across grades. This will require more work, but designing lessons that move students through the crosscutting concept progression while teaching the core ideas and engaging students in the appropriate scientific practices will help ensure that students are doing science in grades K–12.

Developing an understanding of how the *Framework*'s three dimensions relate to the Four Strands of Science Proficiency in *Taking Science to School* is important. Figure 1 (p.58) presents the relationships between the strands and the dimensions. The emerging evidence on science learning from *Taking Science to School* as well as *Ready, Set, Science!* (NRC 2007, 2008) suggests the development of the science proficiencies is best supported when learning environments effectively interweave all four strands into instruction. A similar recommendation from the *Framework* is to interweave the crosscutting concepts and the scientific and engineering practices with the core ideas. What the research tells us is the primary focus for planning and instruction needs to be longer sequences of learning and teaching. The agenda is one of alignment between curriculum-instruction-assessment in classrooms where both teaching and learning is coordinated around "making thinking visible" opportunities employing talk, arguments, models, and representations. Keep this in mind as you read the overviews of the

**Figure 1. Relationship of strands and dimensions (NRC 2012, p. 254)**

| Strands From *Taking Science to School* | Dimensions in *Framework* | How the *Framework* Is Designed to Deliver on the Commitment in the Strand |
|---|---|---|
| 1. Knowing, using, and interpreting scientific explanations of the natural world | Disciplinary core ideas<br><br>Crosscutting concepts | Specify big ideas, not lists of facts:<br><br>Core ideas in the framework are powerful explanatory ideas, not a simple list of facts, that help learners explain important aspects of the natural world.<br><br>Many important ideas in science are crosscutting, and learners should recognize and use these explanatory ideas (e.g., systems) across multiple scientific contexts. |
| 2. Generating and evaluating scientific evidence and explanations<br><br>4. Participating productively in scientific practices and discourse | Practices | Learning is defined as the combination of both knowledge and practice, not separate content and process learning goals.<br><br>Core ideas in the framework are specified not as explanations to be consumed by learners. The performances combine core ideas and practices. The practices include several methods for generating and using evidence to develop, refine, and apply scientific explanations to construct accounts of scientific phenomena. Students learn and demonstrate proficiency with core ideas by engaging in these knowledge-building practices to explain and make scientifically informed decisions about the world. |
| 3. Understanding the nature and development of scientific knowledge | Practices<br><br>Crosscutting concepts | Practices are defined as meaningful engagement with disciplinary practices, not rote procedures:<br><br>Practices are defined as meaningful practices, in which learners are engaged in building, refining, and applying scientific knowledge, to understand the world, and not as rote procedures or a ritualized "scientific method."<br><br>Engaging in the practices requires being guided by understandings about why scientific practices are done as they are—what counts as a good explanation, what counts as scientific evidence, how it differs from other forms of evidence, and so on. These understandings are represented in the nature of the practices and in crosscutting concepts about how scientific knowledge is developed that guide the practices. |

crosscutting concepts in the next section. Ask yourself: How would I integrate the concepts into planning, teaching, and assessing science units?

## The second dimension—seven crosscutting concepts

1. Patterns
2. Cause and effect: Mechanism and explanation
3. Scale, proportion, and quantity
4. Systems and system models
5. Energy and matter: Flows, cycles, and conservation
6. Structure and function
7. Stability and change

Look familiar? The set of crosscutting concepts in the *Framework* is similar to Unifying Concepts and Processes in the *National Science Education Standards* (NRC 1996), Common Themes in *Science for All Americans* (AAAS 1989), and Unifying Concepts in *Science: College Board Standards for College Success* (College Board 2009) (see Figure 2). Regardless of the labels used in these documents, each stresses, like the *Framework,* the importance that "students develop a cumulative, coherent, and usable understanding of science and engineering" (p. 83). The crosscutting concepts are the themes or concepts that bridge the engineering, physical, life, and Earth/space sciences; in this sense they represent knowledge about science or science as a way of knowing. As such, the crosscutting concepts are very important for addressing the science literacy goals.

The first two concepts are "fundamental to the nature of science: that observed *patterns* can be explained and that science investigates *cause-and-effect* relationships by seeking the

**Figure 2. Disciplinary bridging concepts**

| NSES Unifying Concepts | AAAS Common Themes | CB Unifying Concepts |
|---|---|---|
| Systems, Order, and Organization | Systems | Evolution |
| Evidence, Models, and Explanation | Models: Physical, Conceptual, Mathematical | Scale |
| Change, Constancy, and Measurement | Constancy and Change | Equilibrium |
| Evolution and Equilibrium | Constancy | Matter and Energy |
| Form and Function | Stability and Equilibrium, Conservation, Symmetry | Interaction |
| | Patterns of Change | Form and Function |
| | Trends, Cycles, Chaos | Models as Explanations, Evidence, and Representations |
| | Evolution | |
| | Possibilities, Rates, Interactions | |
| | Scale | |

mechanisms that underlie them. The next concept—*scale, proportion, and quantity*—concerns the sizes of things and the mathematical relationships among disparate elements. The next four concepts—*systems and system models, energy and matter, structure and function,* and *stability and change*—are interrelated in that the first is illuminated by the other three. Each concept also stands alone as one that occurs in virtually all areas of science and is an important consideration for engineered systems as well" (NRC 2012, p. 85).

## Progressions for teaching grades K–12

The *Framework* presents each crosscutting concept in two sections, a description followed by a synopsis statement that outlines the developmental features of increasingly sophisticated enactments by pupils. The statements below are from the crosscutting concepts chapter of the *Framework.* The grade band progression descriptions are representative and are not fixed; any one may begin sooner or later according to the development, experiences, and conceptual understandings of the students.

1. ***Patterns.* Observed patterns of forms and events guide organization and classification, and they prompt questions about relationships and the factors that influence them.**
   - K–2 Pattern recognition occurs before children enter school. Develop ways to record patterns they observe. Engage pupils in describing and predicting patterns focusing on similarities and differences of characteristics and attributes.
   - 3–5 Classifications should become more detailed and scientific. Students should begin to analyze patterns in rates of change.
   - 6–8 Students begin to relate patterns to microscopic and atomic-level structures.
   - 9–12 Observe and recognize different patterns occurring at different scales within a system. Classifications at one scale may need revisions at other scales.

2. ***Cause and effect: Mechanism and explanation.* Events have causes, sometimes simple, sometimes multifaceted. A major activity of science is investigating and explaining causal relationships and the mechanisms by which they are mediated. Such mechanisms can then be tested across given contexts and used to predict and explain events in new contexts.**
   - K–2 Children look for and analyze patterns in observations or in quantities of data. Begin to consider what may be causing the patterns.
   - 3–5 Students routinely ask about cause-effect relationships particularly, with unexpected results—how did that happen?
   - 6–8 Engage in argumentation starting from students' own cause-effect explanations and compare to scientific theories that explain causal mechanisms.
   - 9–12 Students argue from evidence when making a causal claim about an observed phenomenon.

3. ***Scale, proportion, and quantity.* In considering phenomena, it is critical to recognize what is relevant at different measures of size, time, and energy and to recognize how changes in scale, proportion, or quantity affect a system's structure or performance.**
   - K–2 Begin with objects, space, and time related to their world using explicit scale models and maps. Discuss relative scales—fastest/slowest—without reference to units of measurement. Begin to recognize proportional relationships with representations of counting, comparisons of amounts, measuring, and ordering of quantities.
   - 3–5 Units of measurement are introduced in the context of length, building to an understanding of standard units. Extend understandings of scale and units to express quantities of weight, time, temperature, and other variables. Explore more sophisticated mathematical representations, e.g., construction and interpretation of data models and graphs.
   - 6–8 Develop an understanding of estimation across scales and contexts. Use estimation in the examination of data. Ask if numerical results are reasonable. Develop a sense of powers of 10 scales and apply to phenomena. Apply algebraic thinking to examine scientific data and predict the effects changing one variable has on another.
   - 9–12 Students acquire abilities to move back and forth between models at various scales and to recognize and apply more complex mathematical and statistical relationships in science.

4. ***Systems and system models.* Defining the system under study—specifying its boundaries and making explicit a model of that system—provides tools for understanding and testing ideas that are applicable throughout science and engineering.**
   - K–2 Express thinking using drawings and diagrams and through written and oral descriptions. Describe objects and organisms by parts; note functions and relationships of parts. Modeling supports clarifying ideas and explanations.
   - 3–5 Create plans; draw and write instructions to build something. Models begin to reveal invisible features of a system—interactions, energy flows, matter transfers. Modeling is a tool for students to gauge their own knowledge.
   - 6–8 Mathematical ideas—ratios, graphs—are used as tools for building models. Align grade-level mathematics to incorporate relationships among variables and some analysis of the patterns therein. Modeling reveals problems or progress in their conceptions of systems.
   - 9–12 Identify assumptions and approximations built into models. Discuss limitations to precision and reliabilities to predictions. Modeling using mathematical relationships provides opportunities to critique models and text and to refine design ideas.

5. ***Energy and matter: Flows, cycles, and conservation.* Tracking fluxes of energy and matter into, out of, and within systems helps one understand the systems' possibilities and limitations.**
   - K–2 Focus is on basic attributes of matter in examining life and Earth systems. Energy is not developed at all at this grade band.
   - 3–5 Macroscopic properties and states of matter, matter flows, and cycles are tracked only in terms of the weights of substances before and after a process occurs. Energy is introduced in general terms only.
   - 6–8 Introduce role of energy transfers with flow of matter. Mass/weight distinctions and idea of atoms and their conservation are taught. Core ideas of matter and energy inform examining systems in life science, Earth and space science, and engineering contexts.
   - 9–12 Fully develop energy transfers. Introduce nuclear substructure and conservation laws for nuclear processes.

6. ***Structure and function.* The way in which an object or living thing is shaped and its substructure determine many of its properties and functions.**
   - K–2 Examine relationships of structure and function in accessible and visible natural and human-built systems. Progress to understandings about the relationships of structure and mechanical functions (wheels, axles, gears).
   - 3–5 Matter has a substructure that is related to properties of materials. Begin study of more complex systems by examining subsystems and the relationships of the parts to their functions.
   - 6–8 Visualize, model, and apply understandings of structure and function to more complex and less easily observable systems and processes. The concept of matter having submicroscopic structures is related to properties of matter.
   - 9–12 Apply the knowledge of structure and function when investigating unfamiliar phenomena; when building something or deciphering how a system works, begin with examining what it is made of and what shapes its parts take.

7. ***Stability and change.* For natural and built systems alike, conditions of stability and determinants of rates of change or evolution of the system are critical elements of study.**
   - K–2 Children arrive at school having explored stability and change. Develop language for these concepts and apply across multiple examples. Help foster asking questions about why change both does and does not happen.
   - 3–5 Explore explanations for regularities of a pattern over time or its variability. A good model for a system should demonstrate how stability and change are related and offer an explanation for both.
   - 6–8 As understanding of matter progresses to the atomic scale, so too should models and explanations of stability and change. Begin to engage in more subtle or conditional situations and the need for feedback to maintain a system.

9–12 Students can model even more complex systems and attend to more subtle issues of stability and change. Examine the construction of historical explanations that account for the way things are today by modeling rates of change and conditions when systems are stable or change gradually, accounting for sudden changes, too.

The message from the *Framework* is that there are important interconnections between crosscutting concepts and disciplinary core ideas. "Students' understandings of these crosscutting concepts should be reinforced by repeated use in the context of instruction in the disciplinary core ideas... the crosscutting concepts can provide a connective structure that supports students' understanding of sciences as disciplines and that facilitates their comprehension of the systems under study in particular disciplines" (p. 101). What this says is that the crosscutting concepts are to be embedded within and conjoined across coherent sequences of science instruction. The *Framework*'s three dimensions—science practices, crosscutting concepts, core ideas—send a clear message that science learning and instruction must not separate the knowing (concepts, ideas) from the doing (practices). Thus, the assessment strategies teachers adopt for pupils' understandings of and enactments with the seven crosscutting concepts must also conjoin the knowing and doing.

## Assessing crosscutting concept learning with learning performances

The *Framework*'s three dimensions represent a more integrated view of science learning that should reflect and encourage science activity that approximates the practices of scientists. What that means for the crosscutting concepts is that assessment tasks should be cumulative across a grade band and contain many of the social and conceptual characteristics of what it means to "do" science; e.g., talk and arguments, modeling and representations. The assessments of crosscutting concepts would be less frequent; each term or annually there would be a performance assessment task that would reveal how students are enacting and using the three dimensions. The majority of assessment tasks for crosscutting concepts will be constructed-response and performance assessments. If the goal is to gauge students' enactments of crosscutting concepts when asked to ascertain patterns, generate mechanisms and explanations, distinguish between stability and change, provide scale representations, model data, and otherwise engage in various aspects of science practices, then the students must show evidence of "doing" science and of critiquing and communicating what was done.

The *Framework* provides teachers with an agreed upon set of curricular goals. The *Next Generation Science Standards* (*NGSS*) assessments are in a "learning performances" format. For example, consider a task to explain how a smell travels through a room. It could be assessed using the grade band information described in section 5, Energy and Matter: Flow, Cycles, and Conservation. The expectation is for students to use some conceptual knowledge (e.g., states of matter) with a practice (e.g., modeling) to develop a mechanism (gas/particle diffusion) that explains the odor's movement. What a teacher is seeking is evidence that students are developing a model of matter made of particles. Related tasks could be mechanisms for the diffusion of a colored dye in water, the separation of sediments in water, or the role of limiting factors in an ecosystem or chemical reaction. The tasks can be gathered over the grade band to develop a portfolio of evidence about students' understandings and enactments of crosscutting concepts.

## Summary

The inclusion of crosscutting concepts in the *Framework* continues a 50-year history in U.S. science education that both scientific knowledge and knowledge about science are important K–12 science education goals. It's the dual agenda for science. The crosscutting concepts are best thought of as the learning goals for science literacy. But success hinges on doing the science. The coordination of the three dimensions reinforces the importance of not separating the doing from the knowing. The alignment of curriculum-instruction-assessment models coordinated around learning progression ideas and research has great potential to organize classrooms and other learning environments around adaptive instruction (targeted feedback to students) and instructed-assisted development. In science over the last century, we have learned how to learn about nature. In education over the last century, we have learned how to learn about learning. As we proceed deeper into the 21st century, let us learn how to meld together these two endeavors. The *Framework* and the *NGSS* are a great beginning, but successful implementation will only come about through the participation and commitment of teachers.

The shift to a "doing" science curriculum focus enacted through the seven crosscutting concepts and the eight scientific and engineering practices will provide students with experiences over weeks, months, and years that will shape their images about the crosscutting concepts, the practices, and, thus, the nature of science. The teacher is the key that will help us unlock how to fully understand the best coherent sequences for learning and teaching.

*Richard A. Duschl* is the Waterbury Chair of Secondary Education at The Pennsylvania State University, and co-chair of the Earth/Space Science writing team for the *NGSS*.

## References

American Association for the Advancement of Science (AAAS). 1989. *Science for all Americans.* New York: Oxford University Press.

College Board. 2009. *Science: College Board standards for college success. http://professionals.collegeboard.com/profdownload/cbscs-science-standards-2009.pdf*

National Research Council (NRC). 2007. *Taking science to school: Learning and teaching science in grades K–8.* Washington, DC: National Academies Press.

National Research Council (NRC). 2008. *Ready, set, science! Putting research to work in K–8 science classrooms.* Washington, DC: National Academies Press.

National Research Council (NRC). 2012. *A framework for K–12 science education: Practices, crosscutting concepts, and core ideas.* Washington, DC: National Academies Press.

# What Does Constructing and Revising Models Look Like in the Science Classroom?

***By Joseph Krajcik and Joi Merritt***

The *Next Generation Science Standards* (*NGSS*) are based on *A Framework for K–12 Science Education* (*Framework*; NRC 2012). The *NGSS* use four key ideas from the *Framework*: (1) a limited number of core ideas of science, (2) the integration or coupling of core ideas and scientific and engineering practices, (3) crosscutting concepts, and (4) the development of the core ideas, scientific practices, and crosscutting concepts over time.

In "Scientific and Engineering Practices in K–12 Classrooms," Rodger Bybee provided an overview of the scientific and engineering practices and showed how they are a refinement and further articulation of what it means to do scientific inquiry in the science classroom.

The *Framework* identifies eight scientific and engineering practices that should be used in science classrooms. These practices reflect the multiple ways in which scientists explore and understand the world and the multiple ways in which engineers solve problems. These practices include:

- Asking questions (for science) and defining problems (for engineering)
- Developing and using models
- Planning and carrying out investigations
- Analyzing and interpreting data
- Using mathematics, information and computer technology, and computational thinking
- Constructing explanations (for science) and designing solutions (for engineering)
- Engaging in argument from evidence
- Obtaining, evaluating, and communicating information

Here, we look in-depth at scientific practice #2—developing, evaluating, and revising scientific models to explain and predict phenomena—and what it means for classroom teaching. Models provide scientists and engineers with tools for thinking, to visualize and make sense of phenomena and experience, or to develop possible solutions to design problems (NRC 2012). Models are external representations of mental concepts. Models can include diagrams, three-dimensional physical structures, computer simulations, mathematical formulations, and analogies. It is challenging for learners to understand that all models only approximate and simplify how the entities they represent work, yet models provide a powerful tool for explaining phenomena. It's critical that a model be consistent with the evidence that exists, and that different models are appropriate in different situations depending on what is being explained. If the model cannot account for the evidence, then the model should be abandoned (Schwarz et al. 2009).

*A Framework for K–12 Science Education* states that by the end of the 12th grade students should be able to:

- Construct drawings or diagrams as representations of events or systems.
- Represent and explain phenomena with multiple types of models and move flexibly between model types when different ones are most useful for different purposes.
- Discuss the limitations and precision of a model as the representation of a system, process, or design and suggest ways in which the model might be improved to better fit available evidence or better reflect a design's specifications. Refine a model in light of empirical evidence or criticism to improve its quality and explanatory power.
- Use (provided) computer simulations or simulations developed with simple simulation tools as a tool for understanding and investigating aspects of a system, particularly those not readily visible to the naked eye.
- Make and use a model to test a design, or aspects of a design, and to compare the effectiveness of different design solutions. (NRC 2012, p. 58).

What does this practice mean for classroom instruction? What does it mean that the practices of modeling will be blended with core ideas? Perhaps the biggest change the modeling practice brings to classroom teaching is the expectation for students to construct and revise models based on new evidence to predict and explain phenomena and to test solutions to various design problems in the context of learning and using core ideas. Students will be engaged in what it means to do science because this is one major activity that drives scientific work and thinking.

Often in science class, students are given the final, canonical scientific model that scientists have developed over numerous years, and little time is spent showing them the evidence for the model or allowing them to construct models that will explain phenomena. As a result, learners often do not see a difference between the scientific model and the phenomena the model is predicting and explaining, or the value of the model for explaining and finding solutions. The *Framework* emphasizes that multiple models might explain a phenomena and that students should improve models to fit new evidence. It is important that science teachers engage students in the modeling process. What do modeling practices look like in the classroom? What are teachers expected to do in their teaching?

It is important for students to construct models that explain phenomena, show how their models are consistent with their evidence, and explain the limitations of those models. Following is one example of what this might look like in a middle school classroom. Imagine a sixth-grade class engaged in exploring core ideas from the *Framework*'s PS1.A: "Gases and liquids are made of molecules or inert atoms that are moving about relative to each other. In a liquid, the molecules are constantly in contact with others; in a gas, they are widely spaced except when they happen to collide." (NRC 2012, p. 108). Blending this core idea with the practice of constructing and revising models, students could be asked to draw a model of how the odor gets from the source to your nose (Merritt and Krajcik 2012; Merritt 2010). Students are asked to complete the task described in Figure 1.

Students are asked to make this model three times during an eight-week unit that focuses on Core Idea PS1.A. In each case, students need to include a key, the drawing, and an explanation of the drawing. Students construct their first model on the first day of the unit. Students walk into class, and the teacher opens a container that contains a strong odor (typically menthol) and asks the students to make a drawing (a representation) of how the odor gets from the container to their noses. The students have had no formal instruction on the particle nature of matter. All they are expected to do is draw a feasible model consistent with the evidence they might see if they had a very powerful instrument that would allow them to "see" the odor.

**Figure 1. Drawing a model of an odor**

Imagine that you have a special instrument that allows you to see what makes up odor. The large circle in the drawing below represents a spot that is magnified many times, so you can see it up close. Create a model of what you would see if you could focus on one tiny spot in the area between the jar and your nose.

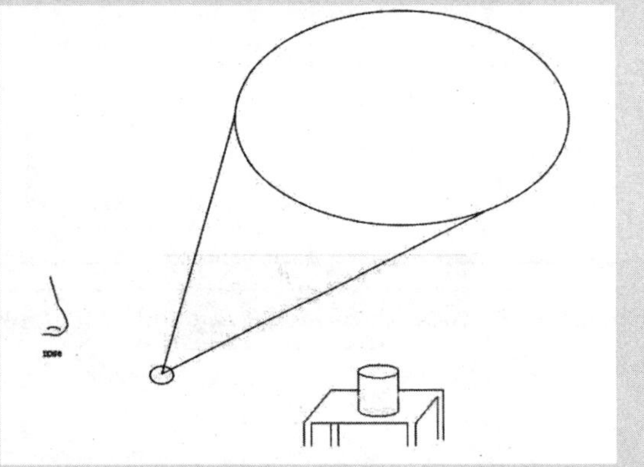

Label the parts of your model, so someone who looks at it will know what the parts represent.

Typically at this initial stage, students' models do not match the scientific model. This is perfectly okay as long as the student model is reasonable and feasible. As previously reported (Novick and Nussbaum 1978), students initially draw a continuous or cloud model to represent the air and the odor. Figure 2 shows an example of what students typically draw.

Next, students complete a series of investigations in which they explore properties of gases. For instance, they use syringes to experience that gases are compressible and expandable: You can add gas to or remove it from a container with a fixed volume without changing the shape of the container. Using these and related experiences, students are again challenged to create a new model of matter to explain how an odor can get from a source to their noses and what they would see if they had a special instrument that "sees" odor. Now, however, their models must be consistent with the evidence they have regarding the properties of gases (i.e., gases can be expanded and compressed and can be added to or taken away from a container with a fixed volume). As Figure 3 (p. 68) shows, students now draw models that are more particulate in nature.

**Figure 2. A student model at the initial stage**

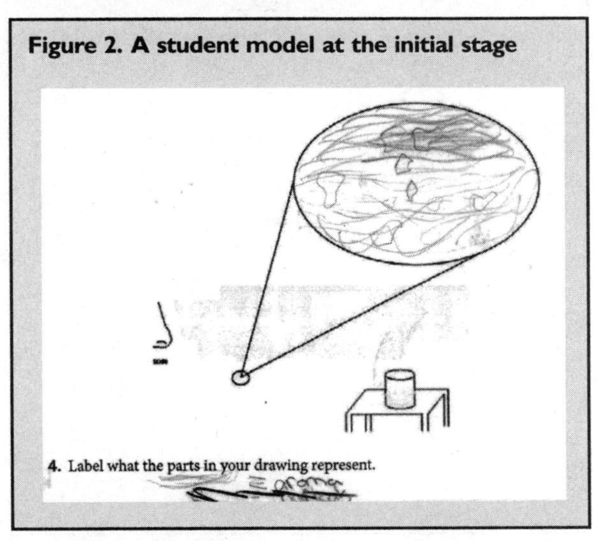


Although this model is still not consistent with the full scientific model, it has features consistent with scientific models. The student now visualizes air and odor to consist of tiny particles too small to see; the particles have space between them and travel in straight lines until they collide with other particles. There are some concerns with the model. For instance, the model shows particles that collide with the imaginary side of the magnified section. The model, however, is consistent with the evidence the student has collected: that a gas can be compressed, expanded, and added to or taken away from a container with a fixed volume.

**Figure 3. A student's second attempt at drawing a model of air and odor**

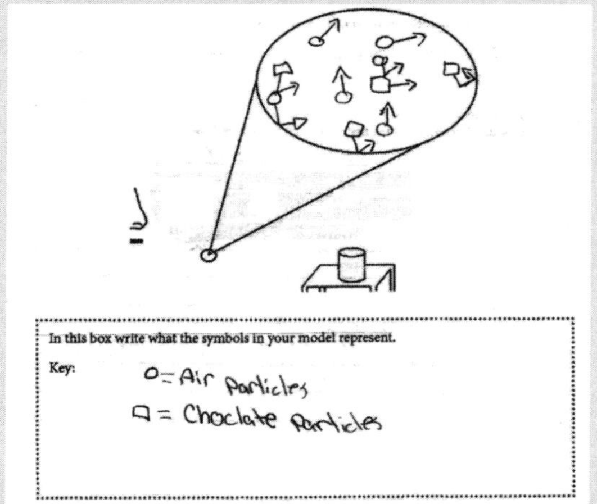

In this box write what the symbols in your model represent.

Key:

O = Air Particles

▱ = Choclate Particles

**Figure 4. A student's model at the end of the unit**

Throughout the unit, students continue to collect additional evidence about the properties of gases. For instance, students explore the effect of temperature on how fast a gas travels by investigating the time it takes ammonia vapor to change indicator paper blue when a test tube containing drops of ammonia is in a warm versus cool water bath. Once students have developed their own models, through careful scaffolding by the teacher, they also develop a class consensus model and explore computer simulations to develop a rich and integrated model of the structure of gases, liquids, and solids as being particulate in nature.

As Figure 4 indicates, at the end of the unit most students have developed models more consistent with the scientific model. The model in Figure 4 shows that gases (air and odor) are made up of tiny particles too small to see, have space between them, move and collide into each other, and change direction as a result of these collisions. There is no indication of the particles colliding with the imaginary walls of the magnified section. Moreover, the student clearly points out there is nothing between the particles. These understandings form a foundation that can be used to build more sophisticated models of the structure of matter. What is important to realize in these examples is that these student models account for all the evidence they have regarding the properties of gases. The student was not told the features of the particle model but rather developed the particle model through carefully supported modeling activities in which students built models based upon evidence. This is the major feature of the modeling practice: developing and revising models.

## Concluding comment

Because *Framework* emphasizes fewer ideas developed across K–12 science curriculum and blended with the use of scientific practices and crosscutting elements, *NGSS* presents a more coherent view of science education that will engage students in the process of doing science.

The U.S. science curriculum has long suffered from being disconnected and presenting too many ideas too superficially, often leaving students with disconnected ideas that cannot be used to solve problems and explain phenomena they encounter in their everyday world. John Dewey expressed this concern in 1910, and we continue to strive so that students learn science in a more coherent manner.

> *Science teaching has suffered because science has been so frequently presented just as so much ready-made knowledge, so much subject-matter of fact and law, rather than as the effective method of inquiry into any subject-matter.* (Dewey 1910)

By focusing on big ideas blended with practices and crosscutting elements over time, the *Framework* and *Next Generation Science Standards* strive to avoid shallow coverage of a large number of topics and allow more time for students to explore and examine ideas in greater depth and use those ideas to understand phenomena they encounter in their lives, while engaging in an "effective method of inquiry." The modeling practices and the example described in this article demonstrate science teaching as an "effective method of inquiry into any subject-matter." This focus on fewer ideas blended with scientific and engineering practices will allow teachers and students time to do science by engaging in a range of scientific practices, including creating and revising models that can explain phenomena and that change as more evidence is collected. Imagine the type of student who emerges from 12th-grade science education after repeatedly experiencing instruction since elementary school that supported them in constructing and revising models to explain phenomena! These students will form a different breed of high school graduates who view science as an "effective method of inquiry" and who will serve as productive 21st-century citizens to create a sustainable planet.

*Joseph Krajcik* is a professor of science education, and *Joi Merritt* is a postdoctoral researcher focusing on science education, both at Michigan State University. Krajcik served as Design Team Lead for the *Framework* and currently serves as Design Team Lead for the *Next Generation Science Standards*.

## References

Dewey, J. 1910. Science as subject matter and method. *Science* 31 (787): 121–127.

Merritt, J. 2010. Tracking students' understanding of the particle nature of matter. Doctoral diss., University of Michigan, Ann Arbor, MI.

Merritt, J., and J. S. Krajcik. 2013. Learning progression developed to support students in building a particle model of matter. In *Structural concepts of matter in science education*, ed. G. Tsaparlis and H. Sevian, pp. 11–45 Dordrecht, Netherlands: Springer.

National Research Council (NRC). 2012. *A framework for K–12 science education: Practices, crosscutting concepts, and core ideas.* Washington, DC: National Academies Press.

Novick, S., and J. Nussbaum. 1978. Junior high school pupils' understanding of the particulate nature of matter: An interview study. *Science Education* 62 (3): 273–281.

Schwarz, C., B. Reiser, E. Davis, L. Kenyon, A. Acher, D. Fortus, Y. Shwartz, B. Hug, and J. S. Krajcik. 2009. Developing a learning progression for scientific modeling: Making scientific modeling accessible and meaningful for learners. *Journal of Research in Science Teaching* 46 (1): 232–254.

# Engaging Students in the Scientific Practices of Explanation and Argumentation

***By Brian J. Reiser, Leema K. Berland, and Lisa Kenyon***

*A Framework for K–12 Science Education* (*Framework*; NRC 2012) identifies eight science and engineering practices for K–12 classrooms. These practices, along with core ideas and crosscutting concepts, define our nation's learning goals for science. An important advance from earlier standards (AAAS 1993, NRC 1996), these practices are clearly identified *not* as separate learning goals that define what students should know *about* the process of science. Instead, the scientific practices identify the reasoning behind, discourse about, and application of the core ideas in science.

The practices outlined in the framework are

- Asking questions and defining problems
- Developing and using models
- Planning and carrying out investigations
- Analyzing and interpreting data
- Using mathematics and computational thinking
- Constructing explanations and designing solutions
- Engaging in argument from evidence
- Obtaining, evaluating, and communicating information

In this article, we examine the sixth and seventh practices concerning explanation and argumentation, respectively. The two practices depend on each other: For students to practice explanation construction, they must also engage in argumentation.

The *Framework* elaborates on how inquiry was expressed in prior standards to add an emphasis on the sensemaking aspects of science (see the earlier chapter by Bybee). The notion of practices moves from viewing science as a set of processes to emphasizing, also, the social interaction and discourse that accompany the building of scientific knowledge in classrooms. This move toward scientific practice requires that we consider the role of argumentation in building knowledge in science because thoughtful and reflective efforts to design investigations, develop models, and construct explanations require critically comparing alternatives, evaluating them, and reaching consensus. In this article, we first define argumentation and explanation individually and then explore their relationship in classroom examples.

## Constructing explanations

The question, "Can you explain that?" is answered in various ways in classrooms. Classroom communities may "explain" by clarifying one's meaning (providing definition), identifying a causal mechanism (explaining why something occurred), or justifying an idea (explaining why one believes the idea) (Braaten and Windschitl 2011). The *Framework* defines explanations as "accounts that link scientific theory with scientific observations or phenomena" (NRC 2012, p. 67), emphasizing that a central form of explanation in science (classroom or professional) is a causal explanation that identifies the underlying chain of cause and effect. This sort of explanation can be evaluated based on whether it can coherently account for—or explain—all of the data students have gathered (NRC 2012, Chapter 3).

The scientific practice of explanation goes beyond defining or describing a named process and links a chain of reasoning to the phenomenon to be explained. So rather than asking students simply to explain cellular respiration, we might ask them to explain *why* a person's exhaled air contains less oxygen than the inhaled air. The explanation should not only describe respiration but also produce a causal chain that fits the evidence that leads to a claim about why oxygen is needed. Such a chain might specify where glucose goes within the body and what materials can enter and exit cells and conclude that a chemical reaction requiring both glucose and oxygen must take place in cells to convert energy to a usable form (NRC 2012, Chapter 9).

In articulating goals for explanation, the *Framework* highlights the process of evaluating ideas to reach the best explanation, including that students should be able to

- use primary or secondary scientific evidence and models to support or refute an explanatory account of a phenomenon; and
- identify gaps or weaknesses in explanatory accounts (their own or those of others).

Thus, developing explanatory accounts includes not only construction but also comparison and critique. Attempts to construct new explanations typically require elements of argumentation to support and challenge potential explanations. Indeed, effective classroom supports for scaffolding explanations reflect these elements of argumentation, such as prompting students to support claims with evidence and reasoning (McNeill and Krajcik 2012; Sutherland et al. 2006). We turn next to unpacking this aspect of scientific practice.

## Engaging in argument from evidence

The practice of arguing from evidence foregrounds the understanding that scientific knowledge is built through "a process of reasoning that requires a scientist to make a justified claim about the world. In response, other scientists attempt to identify the claim's weaknesses and limitations" (NRC 2012, p. 71). This process of scientific argumentation occurs when a claim, perhaps a proposed explanation, is in doubt or is contested (Osborne and Patterson 2011), thereby motivating participants to defend their own and challenge or question alternatives (Berland and Reiser 2009). Chapter 3 of the *Framework* (NRC 2012) teases apart several goals that refer to supporting and contesting knowledge claims:

- Construct a scientific argument showing how the data support the claim.
- Identify possible weaknesses in scientific arguments, appropriate to the students' level of knowledge, and discuss them using reasoning and evidence.

- Identify flaws in their own arguments and modify and improve them in response to criticism.

Scientific knowledge building combines these practices, constructing candidate explanations of natural phenomena and arguing for those claims. As scientists consider alternative interpretations of the same observations, they argue to identify weaknesses in various explanations and incrementally construct a consensus account (possibly drawing elements from multiple sources), arriving at the explanation that best fits the evidence. The interdependence is an example of how the practices interrelate: In response to questions, explanations are developed through analyses of data from investigations and refined through argumentation.

## What makes these practices?

The *Framework* uses "the term 'practices,' instead of a term such as 'skills,' to stress that engaging in scientific inquiry requires coordination both of knowledge and skill simultaneously" (NRC 2012, p. 30). In his chapter earlier in this volume, Bybee emphasizes this expansion of inquiry into the notion of practices to learn "about experiments, data and evidence, social discourse, models and tools," and to engage in using these to "evaluate knowledge claims, conduct empirical investigations, and develop explanations." The practices involve *doing* the work of building knowledge in science and *understanding* why we build, test, evaluate, and refine knowledge as we do. This involves students engaging and reflecting on the practices to develop a sense of *how* the scientific community builds knowledge. This is made explicit in the additional goals that specify that students should be able to explain how and why they engage in argumentation:

- Recognize that the major features of scientific arguments are claims, data, and reasons and distinguish these elements in examples.
- Explain the nature of the controversy in the development of a given scientific idea, describe the debate that surrounded its inception, and indicate why one particular theory succeeded.
- Explain how claims to knowledge are judged by the scientific community today and articulate the merits and limitations of peer review and the need for independent replication of critical investigations.

Developing these understandings of scientific knowledge building requires adopting the goals of these practices. If we expect students to learn that the scientific community builds knowledge by constructing explanations and arguments, then they must experience using these practices to address questions they have identified. Furthermore, the student participation must be *meaningful*, so that students argue to resolve inconsistencies in their explanations and not because their teacher asked them to (Berland and Reiser 2009).

We illustrate this idea of meaningful engagement in explanation and argumentation through four classroom examples.

### Example 1: Arguing for predictions strengthens explanations

In the first example (Hammer and van Zee 2006), we see that students encouraged to defend their predictions constructed causal explanations about why differently shaped objects fall to the ground at different rates (Core Ideas PS2.A and PS2.B). On the first day of this investigation, first-grade students and their teacher worked to explain what happened when they dropped a sheet of a paper and a book. They concluded that the book falls first because it has "more strength" (the students' word for weight). This discussion also introduced ideas related to gravity and wind resistance. On the second day, the class predicted what would happen if they dropped a book and a crumpled piece of paper. Brianna predicted that "They will fall at the same time 'cause they both got the same strength together." This idea aligned with other student suggestions that the crumpled paper "weighed more" than the original paper. When the teacher questioned how the weight could have changed, Rachel added to Brianna's idea, saying, "The paper ... used to be, um, really light ... but [now] it probably has as much strength as the book since all the, um, paper is crumpled up together."

After a pause, Brianna said, "If it's balled up, it's still not heavy, it's the same size." Then Brianna questioned her own explanation and pushed the class to reconsider their assumption that the paper weighed more when crumpled. Numerous students said they agree that crumpling paper wouldn't change its weight. Diamond then said, "The first time, like this [flat], and then it balled up." In other words, the paper changed shape. Diamond added that the crumpled paper did not drift to the ground as the flat sheet did. As Brianna stated, "It just drops, kind of like the book." While there is still important work to be done to tease apart shape and weight in the discussion, the example demonstrates how defending (or arguing for) predictions by explaining why the event occurred enabled students to investigate and question their initial assumptions about the paper and the relationship between the paper's shape and its fall to the ground.

### Example 2: Reconciling competing explanations

In the second example, students develop explanations to defend predictions, as in Example 1, but also reconcile their differences, helping them move toward a more scientifically accurate understanding. In this case, a mixed-grade classroom of fifth- and sixth-grade students investigated how tectonic plates move and interact (Core Idea ESS2.B, 6-8). Before this investigation, students discussed convection currents and constructed models of particular plate boundaries: A third of the class modeled convergent boundaries, a third focused on divergent boundaries, and a third on transform boundaries. On the third day, students formed groups aligned with the three types of boundaries to explore a question that emerged: With all this plate motion, is the Earth staying the same size or getting bigger or smaller?

In one group of four students, two believed the Earth was staying the same size and two thought it was getting bigger. Pint argued that dinosaur fossils "prove" that the Earth is getting bigger because they are evidence that the Earth's layers are getting thicker.

***Pint*:*** You have to dig and dig [to find the dinosaur bones]. So that means the Earth has been getting larger because you have to dig so much to get to bones... (1)

***Olive:*** Yeah, we saw *Jurassic Park*, I guess. (2)

*Intervening additional discussion of dinosaur fossils and teacher interruption.*

***Fern:*** I understand how you think of the dinosaur bones. But those are convergent that have covered the dinosaur bones. But not all convergence makes mountains. Some meet [gestures that plates meet and stay flat]. So the dinosaur bone was one plate and then that plate kind of moved and then that converged and overlapped. (18)

*Intervening discussion of whether convergent boundaries always create mountains.*

***Fern:*** So let's say some dirt moved over here, but then there's some dirt not over there. There still might be dirt over there. So it's still even because that dirt over here came from over there. So the world is even, and it's not growing, because the magma might come in but then it diverges and collapses. (24)

*Students selected their own pseudonyms

This episode illustrates the relationship between argumentation and explanation when students engage meaningfully in the practices. This happens when students actively listen and respond to one another. For example, in line 2 Olive connects her own experiences with Pint's points. Fern, in line 18, similarly addresses her teammates' ideas about the explanation they are building: "I understand how you think of the dinosaur bones…" (Line 18). Fern then uses the language and imagery of Pint's understanding—that of dinosaur bones proving that the Earth's plates are layering on top of one another—to move the conversation toward her own (more scientifically accurate) understanding, that the Earth "is still even" (the same size). Fern states,"Let's say some dirt moved over here but then there's some dirt not over there … so it's still even because that dirt over here came from over there" (Line 24).

The spontaneity of the students' discourse—they are not looking at a worksheet or obviously thinking about their teacher's expectations—suggests that these interactions are meaningful. The students are actively engaged in figuring this out—in constructing an explanation regarding whether and how the plate motion affects the shape and size of the Earth. A less meaningful engagement is easy to imagine—the students could have been given a worksheet that asked for evidence: "This says we need to find evidence for our idea." Alternatively, they could all have worked to answer the question individually without much cross-talk or requested that their teacher tell them the answer to the question. Instead, however, they are engaged in what appears to be purposeful knowledge-construction interactions.

This interaction provides evidence of both explanatory and argumentative practices. Students work to construct an explanation of how tectonic plate movement affects the shape and size of the Earth. For example, in line 24 Fern offers an explanation regarding *how* the tectonic plates could move without changing the overall size of the Earth. Together students reason how this could occur and also explain Pint's observation that dinosaur bones are "buried." The argumentative nature of the discussion is apparent when they engage in nascent forms of the first two argumentative goals in the *Framework,* justifying their own ideas (lines 1 and 24) and challenging alternative ideas (as Fern challenges Pint).

### Example 3: Building consensus from multiple contributions

In this third example, fifth-grade students use their ideas to defend, make sense, and build a consensus. The students investigated condensation (Kenyon, Schwarz, and Hug 2008) and represented their explanations for how water appeared on a cold pop can in a diagrammatic model focusing on changes of state. The goal of the unit was an initial form of the particle model (Core Idea PS1.A, 3-5), in which the existence of water as particles in gas state can explain where the water comes from in condensation and where it goes in evaporation. The day before this discussion, students in the group evaluated each other's individual models. Here they construct a group consensus model of condensation using ideas from those individual models. (In this classroom, the teacher extended the targeted PS1.A, 3-5 learning goal to also bring in kinetic energy, Core Idea PS3.A, 6-8, as part of the explanation).

*Amy:* Wait guys! Why do we think why condensation shows up? Can anybody? (1)

*Amy:* Yeah, but why do you think it got there? Because of the water in the air? (2)

*Jenny:* Because, of the temperature… (3)

*Amy:* The coldness is taking the kinetic energy from the air…. (4)

*Ivan:* Coldness isn't a word! (5)

*Amy:* Okay, does everyone agree that the kinetic energy is taking away from air and turning it into a liquid? (6)

*Ivan:* Sure! (7)

*Amy:* We should write when gas loses kinetic energy (KE), it turns into a liquid, and when liquid loses kinetic energy it turns into a solid. Or we could write gases minus KE of the liquid. Liquid minus KE equals… (8)

*Ivan:* So what are we doing? (9)

*Amy:* Explanations! (10)

*Mary:* Condensation always occurs on the surface that is cooler than the air. (11)

*Jenny:* Okay! (12)

*Ivan:* Condensation works when the water vapor loses its KE and turns into a liquid. (13)

*Amy/Jenny:* That's what we said! (14)

*Ivan:* I know! (15)

*Matthew:* We can't say condensation ALWAYS OCCURS! (16)

*Lori:* … always occurs on COLD surfaces! (17)

| | |
|---|---|
| ***Matthew:*** | What if you're in a spot that has no humidity whatsoever? (18) |
| ***Jenny:*** | Mr. Smith explained that with the warm can in front of the humidifier, nothing happens. (19) |
| ***Matthew:*** | Okay. (20) |

The conversation, by attempting to fill gaps, involves several important reformulations that clarify the overall explanation. "Water in the air," "temperature," and, later, "kinetic energy" emerge as important steps in the mechanism. Three attempts to formulate what they have figured out (lines 8, 11, 13) lead to Ivan's summary that references water vapor losing kinetic energy and turning to a liquid. Matthew raises a final concern (lines 16, 18) to clarify the conditions under which condensation occurs. In response, important additional qualifications are added by Lori (line 17, the surface must be colder than air) and Jenny (line 19, there must be sufficient water in the air). Consequently, Matthew's concern is met by modifying Ivan's proposed explanation to produce a new synthesis, which is reflected in the group's articulated model (See Figure 1).

**Figure 1. A group's articulated model of condensation on a soda can**

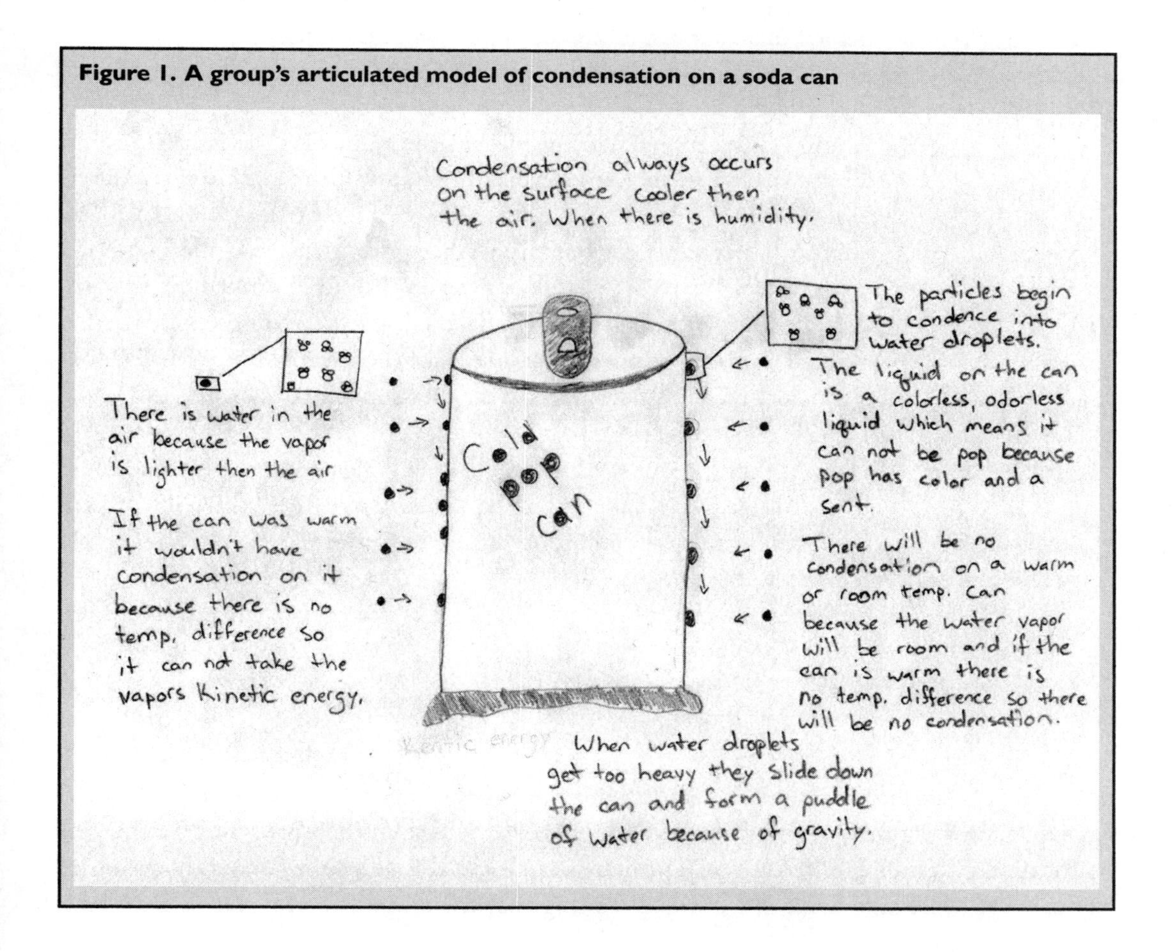

### Example 4: Critique leads to clarified explanation

The final example demonstrates how critiques can lead students to improve and clarify their explanations. In this case, a group of eighth graders undertook an investigation of population change (Core Idea LS4.B). Students shared ideas to account for changes in populations of Galápagos finches over time. Students discovered that during a drought, most of the birds died, and they attempted to explain why the birds died and why others survived (Tabak and Reiser 2008).

***Mr. N:*** So, I get where Ina is going. Can someone put this in other words? What do you mean "the populations adapt"? Ina, you have the floor so you can call, or you can keep pushing on this if you want. (1)

[Ina calls on Joe] (2)

***Joe:*** Well, like the beak length thing. The reason why there were so many birds that had bigger beak lengths in '77 and '78 was well, the ones with the bigger beaks survived, and they mated, so their babies, they had the trait, the bigger beaks. And that's why most of their babies and the adults had bigger beaks rather than medium or small beaks because that trait helped them survive in the drought. [Joe calls on Jeff] (3)

***Jeff:*** The reason the population adapts is to survive. If they don't adapt then they will die so they'll disappear, so yeah... Kelly? (4)

***Kelly:*** Umm, I think it's because the birds with the smaller beaks died, and the longer beaks were able to have children, and their children had longer beaks, so they survived and the trait was being passed on a lot. Ina? (5)

***Ina:*** Umm, I don't think so. Because we have this graph that shows the wet [season] of 1973 to the dry [season] of 1978, and it jumped up. It wasn't that the ones with the shorter beaks died. Even the longest beak here is like pretty much even with the middle of the pack in 1978. Mr. N? (6)

***Mr N:*** So you're saying it's not always every short one dies? (7)

***Ina:*** Yeah. (8)

***Mr. N:*** Okay. Is that true for the moths too? Was it always every peppered moth dies? [The students had earlier explained why some variations of peppered moths survived pollution during the late 1800s.]

***Most students:*** No. (9)

***Mr. N:*** Just, even for the moths, it's kind of like the odds change some, right? (10)

***Most students:*** Yeah. (11)

***Mr. N:*** Okay, so I think I get what you're arguing. (12)

This episode shares important aspects with the previous episodes. Like example 2, there are explanatory accounts proposed and a critique raised. A resolution is proposed (in this case by the teacher bringing in features of a prior explanation). This keeps the core of the proposed explanation while addressing the critique (adding that the advantage of a trait is like "odds changing" rather than "always every" bird or moth lacking the trait dies. Although this excerpt does not go as far toward having the students articulate the consensus that resolves the critique, the class agrees with the teachers' proposed change that handles Ina's concern (line 6), while managing to retain the central parts of the causal chain proposed by Joe (line 3) and Kelly (line 5). In this short excerpt, the students developed a logical chain that reflects some of the most important steps in natural selection: preexisting variation of a trait (beak length), changing environmental conditions ("the drought"), differential survival ("the ones with the shorter beaks died"), and heritability of the trait ("passed on a lot"). (Missing from the account is an explanation for why birds with longer beaks were more likely to survive.)

## Conclusions

Across the four examples, we see that students arguing for their explanations can strengthen those explanations and help construct a consensus explanation. We see this in examples 1 and 3, in which the support, defense, and consensus building helped make the explanations more elaborate and precise; and in examples 2 and 4, in which this argumentation made the explanations better able to handle possible contradictions. In this way, the explanations improve along several of the dimensions outlined in the *Framework,* improving the causal account (filling gaps) and articulating and improving their fit with evidence.

In addition, in each of these examples the students engaged in meaningful forms of scientific practices—they were working to make sense of scientific phenomena rather than working to replicate the understandings communicated by a textbook or other authority. These examples illustrate student engagement in the practices of science rather than in the processes or skills of science. Together, these examples illustrate the importance of considering how the scientific practice of argumentation plays a role in bringing explanations into K–12 classrooms.

These examples and related research suggest how classroom environments might support this meaningful engagement in scientific practice. We, as educators, must create situations that enable students to interpret the practices of explanation and argumentation as something they could reasonably do to construct knowledge (Berland and Hammer 2012). This requires focusing on reasons for ideas, rather than only on the accuracy of a particular idea (Sutherland et al. 2006). It requires creating a climate that is safe for students to be wrong as they work toward more complete explanations. It also requires asking students rich questions that have multiple plausible answers so that students can discuss and reconcile them, developing consensus explanations.

*Brian J. Reiser* is a professor of learning sciences at Northwestern University in Evanston, Illinois. Reiser served on the team that developed the NRC 2012 *A Framework for K–12 Science Education Standards*. *Leema K. Berland* is an assistant professor of STEM education at the University of Texas in Austin, Texas. *Lisa Kenyon* is an associate professor in the departments of biological sciences and teacher education at Wright State University in Dayton, Ohio.

## References

American Association for the Advancement of Science (AAAS). 1993. *Benchmarks for science literacy.* New York: Oxford University Press.

Berland, L. K., and D. Hammer. 2012. Framing for scientific argumentation. *Journal of Research in Science Teaching* 49 (1): 68–94.

Berland, L. K., and B. J. Reiser. 2009. Making sense of argumentation and explanation. *Science Education* 93 (1): 26–55.

Braaten, M., and M. Windschitl. 2011. Working toward a stronger conceptualization of scientific explanation for science education. *Science Education* 95 (4): 639–669.

Hammer, D., and E. H. van Zee. 2006. *Seeing the science in children's thinking: Case studies of student inquiry in physical science.* Portsmouth, NH: Heinemann.

Kenyon, L., C. Schwarz, and B. Hug. 2008. The benefits of scientific modeling. *Science and Children* 46 (2): 40–44.

McNeill, K. L., and J. Krajcik. 2012. *Supporting grade 5–8 students in constructing explanations in science: The claim, evidence, and reasoning framework for talk and writing.* New York: Allyn and Bacon.

National Research Council (NRC). 1996. *National science education standards.* Washington, DC: National Academies Press.

National Research Council (NRC). 2012. *A framework for K–12 science education: Practices, crosscutting concepts, and core ideas.* Washington, DC: National Academies Press.

Osborne, J. F., and A. Patterson. 2011. Scientific argument and explanation: A necessary distinction? *Science Education* 95 (4): 627–638.

Sutherland, L. M., K. L. McNeill, J. S. Krajcik, and K. Colson. 2006. Supporting middle school students in developing scientific explanations. In *Linking science and literacy in the K–8 classroom,* ed. R. Douglas and K. Worth, pp. 163–181. Arlington, VA: NSTA Press.

Tabak, I., and B. J. Reiser. 2008. Software-realized inquiry support for cultivating a disciplinary stance. *Pragmatics and Cognition* 16 (2): 307–355.

# Obtaining, Evaluating, and Communicating Information

*By Philip Bell, Leah Bricker, Carrie Tzou, Tiffany Lee, and Katie Van Horne*

The National Research Council's recent publication *A Framework for K–12 Science Education: Practices, Crosscutting Concepts, and Core Ideas* (*Framework*; NRC 2012), which is the foundation for the *Next Generation Science Standards*, places unprecedented focus on the practices involved in doing scientific and engineering work. In an effort to lend specificity to the broad notion of "inquiry," the intent behind the practices outlined in the *Framework* is for students to engage in sensible versions of the actual cognitive, social, and material work that scientists do. This article focuses on one of those practices.

## Obtaining, evaluating, and communicating information

Reading and writing comprise over half of the work of scientists and engineers (NRC 2012; Tenopir and King 2004). This includes the production of various scientific representations—such as tables, graphs, and diagrams—as well as other forms of communication such as giving conference presentations and speaking to the public and other stakeholders. The reading and writing that scientists do help them better understand scientific ideas and communicate their research to their colleagues and to the public. Thus, K–12 students of science should have substantial and varied experiences with reading, analyzing, writing, and otherwise communicating science so that they too can deeply engage with disciplinary core ideas and crosscutting concepts while exploring practices associated with scientific reading and writing. This is why the "obtaining, evaluating, and communicating information" practice was included in the *Framework.*

K–12 students should learn how to conceptualize, compose, and refine different types of scientific writing from detailed scientific research abstracts to articles for a lay audience on current issues related to topics such as health and the environment to elaborate evidence-based arguments and even to proposals for funding. They should also learn how to find and understand everything from science-related newspaper articles to peer-reviewed journal articles—at reading levels that are developmentally appropriate and with use of relevant disciplinary criteria to select pieces and judge their quality. K–12 students also need practice obtaining information and evaluating it (to make personal health decisions or take informed action on environmental issues, for example). Students should learn to search and browse scientific and library databases, the internet, and print and digital media outlets (newspapers, magazines, blogs, Twitter, RSS feeds) for information they can use to inform their research and learning of science. They need to practice evaluating the information they find, learning how to judge whether information is credible and by whose criteria, as well as learning which information is necessary and useful for any given purpose.

In articulating the related learning goals, the *Framework* (NRC 2012, pp. 75–76) specifies that all students should be able to:

- Use words, tables, diagrams, and graphs, as well as mathematical expressions, to communicate their understanding or to ask questions about a system under study.
- Read scientific and engineering text, including tables, diagrams, and graphs, commensurate with their scientific knowledge and explain the key ideas being communicated.
- Recognize the major features of scientific and engineering writing and speaking and be able to produce written and illustrated text and oral presentations that communicate their own ideas and accomplishments.
- Engage in a critical reading of primary scientific literature (adapted for classroom use as appropriate) and of media reports of science and discuss the validity and reliability of associated data, hypotheses, and conclusions.

## Instruction as a "cascade of practices"

The *Framework* calls for students to routinely participate in extended science and engineering investigations that engage them in authentic practices while learning about disciplinary core ideas and making connections to the crosscutting concepts. Direct participation in scientific and engineering work will support students' science learning and the scientific literacy goals of the *Framework*. We argue that it will also help students understand specific career possibilities in the sciences and in engineering.

The practices do not operate in isolation, and we argue that part of giving students opportunities to participate in authentic scientific and engineering work is ensuring that they can experience firsthand the interrelatedness of these practices—as an unfolding and often overlapping sequence, or a cascade. For example, students may begin by learning about natural resources and posing a testable scientific question (practice 1) before designing a study and collecting data (practice 3), analyzing and interpreting those data (practice 4), developing a model (practice 2), and communicating important aspects of that model to an audience (practice 8). Many such permutations exist for sequencing and overlapping the practices during investigations, depending on the type of scientific or engineering investigation underway and the specific learning goals in question.

## Promoting educational equity through practices

The focus on practices can also advance an educational equity agenda. There is often an artificial distinction made in science learning experiences between what counts as science and what is not science (Calabrese Barton 1998; Warren et al. 2003). Removing this barrier allows for learners to make connections between their lives and science and engineering and allows for diverse voices to be heard (Calabrese Barton 1998, p. 389). This is particularly important for the language-intensive practice of obtaining, evaluating, and communicating information. The *Framework* describes another instructional strategy: "Recognizing that language and discourse patterns vary across culturally diverse groups, researchers point to the importance of accepting, even encouraging, students' classroom use of informal or native language and familiar modes of interaction" (NRC 2012, p. 285). These inclusive instructional strategies allow students to leverage what they know and participate in the workof science focused on community interests and practices.

### Example 1 (Prekindergarten): Beginning a science research practice with our youngest students

Young children are curious about the world around them and readily engage in informal science throughout their everyday lives. The *Framework* calls for a significant focus on providing science learning opportunities in preschool and early elementary school, so it is important to consider how young students still learning to read and write can engage in the practices of science. Through a multiyear research collaboration with two prekindergarten classrooms, the team has developed an approach to science instruction that aligns to the vision in the *Framework* by incorporating students' science-related interests and experiences while engaging them in practices, developing an understanding of core ideas, and making connections to crosscutting concepts.

During a unit early in the school year, one teacher was reflecting on all of the questions her students had been asking about the natural world and their varied interests related to the unit. Realizing that she did not have enough time to address each student's individual questions, she came up with an activity that became known as "Research Day." Students were given classroom time to do their own research using relevant nonfiction books preselected by the teachers and the school librarian, and then they drew, dictated, and shared their research findings with their peers.

In a later unit about garden ecosystems, students asked many questions about insects and other living creatures found in a garden (e.g., aphids, bees, worms, spiders, etc.), so the teachers offered another Research Day. One student, Eleanor, was immediately attracted to a book with colorful illustrations of ladybugs in a garden. A teacher came over to read the text to her, and Eleanor, satisfied with her book selection, drew a detailed picture of a ladybug surrounded by aphids on her research paper. She then dictated information about ladybugs to be written on her paper by a teacher: "Sometimes ladybugs' food runs out, and there are not enough aphids to go around. The ladybugs gather in a swarm and fly off somewhere near to survive." Here, the teacher's support of the students' individual interests allowed Eleanor to find information that provided further evidence related to Core Idea LS2 (Ecosystems: Interactions, Energy, and Dynamics) in their garden ecosystems unit: Animals depend on their surroundings for survival.

All students were given time to look through books and document their newly learned information through drawings and dictations, just like Eleanor. At the end of Research Day, the students stood in front of the class to share their research papers with their peers, describing their drawings and explaining what they learned that day. Later, the teachers compiled the research papers into a book that was displayed in the classroom. Research Day was repeated during various units throughout the year, resulting in a collection of student research that was reviewed by the students and their parents.

### Example 2 (Grade 5): Using public service announcements to communicate the science behind everyday health practices

The *Micros and Me* curriculum unit focuses students on the learning of microbiology by connecting it to personally and community-relevant health issues. We incorporate inquiry investigations, such as investigating the presence of beneficial microorganisms such as yeast, sampling for microorganisms in school, and conducting student-centered investigations about hand washing and "green" cleaning. The design has two goals: (1) making science personally consequential to students' lives, and (2) connecting authentic scientific practices and content deeply with students' everyday practices. Students learn about the characteristics of life such as reproduction (LS1.B) and the structure of plant and animal cells (LS1.A). While learning about growth of "micros" (bacteria, viruses, fungi), they learn that organisms have certain requirements for life (LS1.C).

One of the central innovations in the curriculum is a *self-documentation* technique (Tzou and Bell 2010) accompanied by community-based interviews conducted by the students to elicit students' family and community-based activities related to health and illness prevention. Self-documentation is a technique where, in this case, students were given digital cameras to take home for one night to document the activities in their lives related to an open-ended prompt. However, we have also used self-documentation in other contexts where students just record in a journal or on a worksheet the activities in their everyday lives related to a prompt.

In *Micros and Me,* students investigate the following prompt: "What are ways that you and your family/community stay healthy and keep from getting sick?" We argue that because non-Western customs and ways of thinking are typically marginalized in traditional school science curricula (Ballenger and Carpenter 2004), it is particularly important—when thinking about broadening participation in science—to find ways to connect a broader range of practices to important curricular goals in science education. In *Micros and Me,* the self-documented activities are connected to a student-led research project where students synthesize information from scientific investigations in the unit, self-document home and community activities, and conduct independent internet and library research on health issues found in their community to construct an evidence-based argument in the form of a *public service announcement,* several of which are displayed in the school and the local public library.

The goal of the public service announcement is threefold: (1) to validate and leverage students' everyday activities within the context of formal science instruction, (2) to give students practice unpacking and evaluating internet and book-based research sources, and (3) to engage students in communication of scientific ideas to a public audience of their choosing. Students are asked to choose a personally relevant health activity to research (e.g., managing asthma), find at least three sources about that activity, and construct a *convincing* public service announcement aimed at persuading their friends and families to take some type of action related to the activity in question. In a public service announcement poster about *E. coli,* written in crayon, we see evidence of the student communicating scientific information in the language that is appropriate to his peer audience. The student gives four examples for avoiding the contraction of *E. coli*: ordering well-cooked meat in a restaurant, not drinking water in lakes, drinking pasteurized juice, and washing hands after using the restroom. Finally, the student translates this information into a list in Spanish on the left side of the poster since that language is prominent in his community. This example shows how empirical and research-focused activities can be integrated with high personal and community relevance by designing instruction to include the communication practice.

### Example 3 (Grade 8): Evaluating and arguing with evidence in a classroom science debate

The third example comes from a curriculum intervention study conducted in an eighth-grade physical science classroom where the teacher made extensive use of computer learning environments to support students' science investigations (Linn and Hsi 2000). This example highlights how two scientific practices—"obtaining, evaluating, and communicating information" and "engaging in argument from evidence"—can be productively sequenced to support students' conceptual learning.

It can be very productive to view science classrooms as "scientific communities writ small" where students produce, share, debate, and refine knowledge in similar ways to how practicing scientists do it. In this unit, students evaluated disparate sources of information—from their classroom experiments, various web sources and advertisements, to their own life experiences—according to scientific criteria. They identified and evaluated this information as they prepared for a classroom debate. The goal of the classroom debate is to come to a group consensus about the topic as a "scientific community."

After conducting four weeks of experiments related to the properties of light embedded in Core Ideas PS4.B and PS4.C (e.g., light intensity over distance, how light travels through space from distant stars, reflection, absorption/energy conversion), students then engaged in an eight-day debate project as a culminating activity for the light unit. They evaluated a shared corpus of evidence, searched out new evidence on the internet, developed detailed written, evidence-based arguments, and engaged in two days of whole class debate about "How Far Does Light Go?" (Bell 2004).

Figure 1 shows the kind of written arguments students authored, for various pieces of evidence in the corpus, when they were given the sentence-starter "We think this supports the theory ____ because…." In addition to this "causal prompt" scaffold, students also reflected on multiple relevant criteria related to how well the evidence fits with scientific knowledge, whether appropriate methods were used, the trustworthiness of the source,

**Figure 1. Two students who analyzed evidence from a shared corpus wrote this explanation**

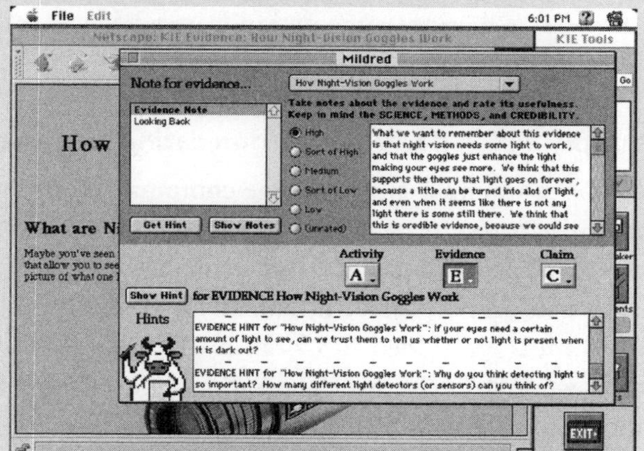

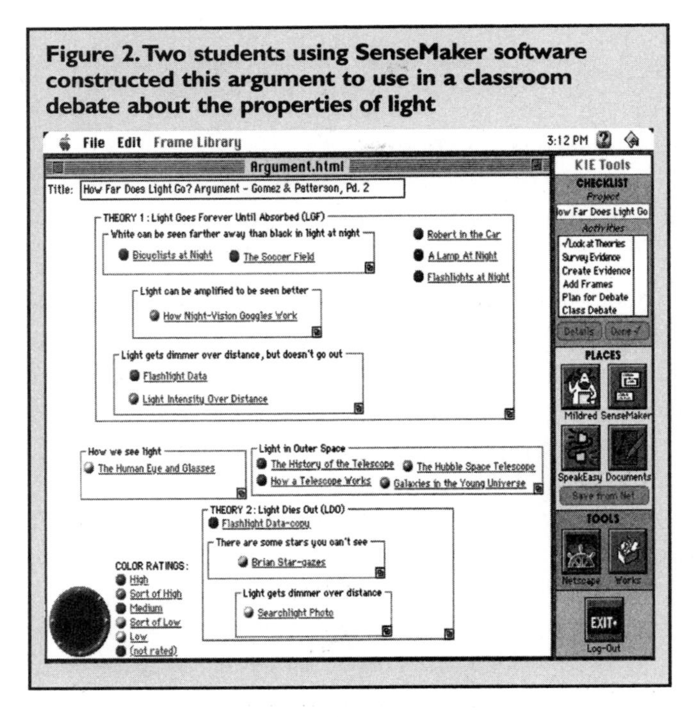


**Figure 2. Two students using SenseMaker software constructed this argument to use in a classroom debate about the properties of light**

and the usefulness of the information for the debate topic. As shown in Figure 2, each pair of students created an argument map using a software tool called SenseMaker that related pieces of evidence (shown as dots) to conceptual claims (shown as boxes). These argument maps allowed for an easy comparison of students' ideas during the classroom debate. The transcript (Figure 3, opposite) highlights the kind of sense-making discussions that happened as students tried to develop a shared understanding of the physics of light.

Student 2 explains the decision to consider a certain phenomenon labeled "The Soccer Field" irrelevant, meaning that it doesn't provide any evidence that can be used to distinguish between the two alternative theories. Student 3 provides a different perspective, saying that the light is stopped at different distances, which leads student 2 to reconsider the evidence. This approach to drawing the relationships between theories and evidence allows for more focused questions to be posed to peers, and the detailed written arguments allowed students to share and refine their conceptual ideas at a deeper level (Figure 3).

Whole class sense-making conversations like this one were shown to support students' conceptual learning about light on cognitive assessments (Bell 2004). Students also developed epistemic knowledge that science is a social enterprise that progresses through the evaluation of evidence, systematic argumentation from evidence, and the collaborative debate of ideas (Bell and Linn 2002).

### Example 4 (Grade 10): Communicating research investigations to scientists

This fourth example showcases the communicative practices of high school biology students who participated in contemporary infectious disease-related research. Students learned the biology behind why various pathogens make humans sick at the cellular level, as well as the science behind how and why infectious diseases are transmitted locally and globally. They learned ideas embedded in Core Idea LS1 (From Molecules to Organisms: Structures and Processes), such as cell structure and function related to the immune system, as well as ideas embedded in Core Idea LS4 (Biological Evolution: Unity and Diversity), such as viral evolution. Students had their choice of project: a local social network analysis in order to learn about and apply constructs like herd immunity or a global epidemic modeling study in order to think about the various factors affecting the spread of infectious disease, such as seasonality and viral latency periods. As part of these projects, students read original research, communicated with scientists who conduct this type of research, and conducted their own research. Students developed products to communicate various aspects of their work to scientists and other health professionals, their teachers, and their peers. These products included: (a) a *research design plan*, (b) an *elevator speech*, and (c) an *original research paper*.

| **Figure 3. Example 3, transcript segment** |
|---|
| [Student 1]<br>Why did you put *The Soccer Field* in **Irrelevant**? |
| [Student 2, presenting to the class]<br>I put *The Soccer Field* in **Irrelevant** because . . . oh yeah—because it was the one with the flashlight and they held the light back and then the light from the car—like headlights they—it went further so it didn't—I don't think it really made a difference. Or I don't think it really supported either theory because it did go a long ways, but the light intensity wasn't as strong. |
| [Student 3]<br>For *The Soccer Field,* doesn't that kind of prove how far light keeps going if it keeps showing as its—as [the guy] keeps moving back and the light—light gets stopped like a reflection or would it stop that light because <UNCLEAR>. |
| [Student 2]<br>Well, I don't think it really supports either theory because I know that the light is still there, and it's being absorbed and it's spreading out so much that you can't see it, but the light energy is still there. |

Once students selected a project, they designed a research study to conduct. Part of this involved reading published social network analysis studies involving infectious diseases or published global epidemic modeling studies (depending on students' project choice), reading background information on analysis and modeling tools, and reading background information on the disease(s) they wanted to use as a case study. Students then wrote their *research design plan* (see Figure 4, p. 88 for an example with expert feedback) where they developed the specifics of the study they wanted to conduct, including their testable question, their rationale(s) for posing that question, their hypothesis, their methods, and their thinking about how they would know if their data supported or refuted their hypothesis (and spoke to their testable question). Once students designed their studies, they forwarded their research plans to scientists and health professionals, who provided feedback (e.g., questions to ponder, challenges to students' thinking, resources to investigate, and lessons learned from their own research). Students then revised their plans based on the feedback and proceeded with their studies.

After students collected and analyzed their data, they wrote *elevator speeches* (Figure 5, p. 88) in which they clearly and succinctly explained the details of their study, including their preliminary findings. They received feedback on the text of their speeches from peers, and they then revised their speeches in preparation for a two-minute presentation to scientists and health professionals. Students answered questions based on their research and the ideas they learned in class.

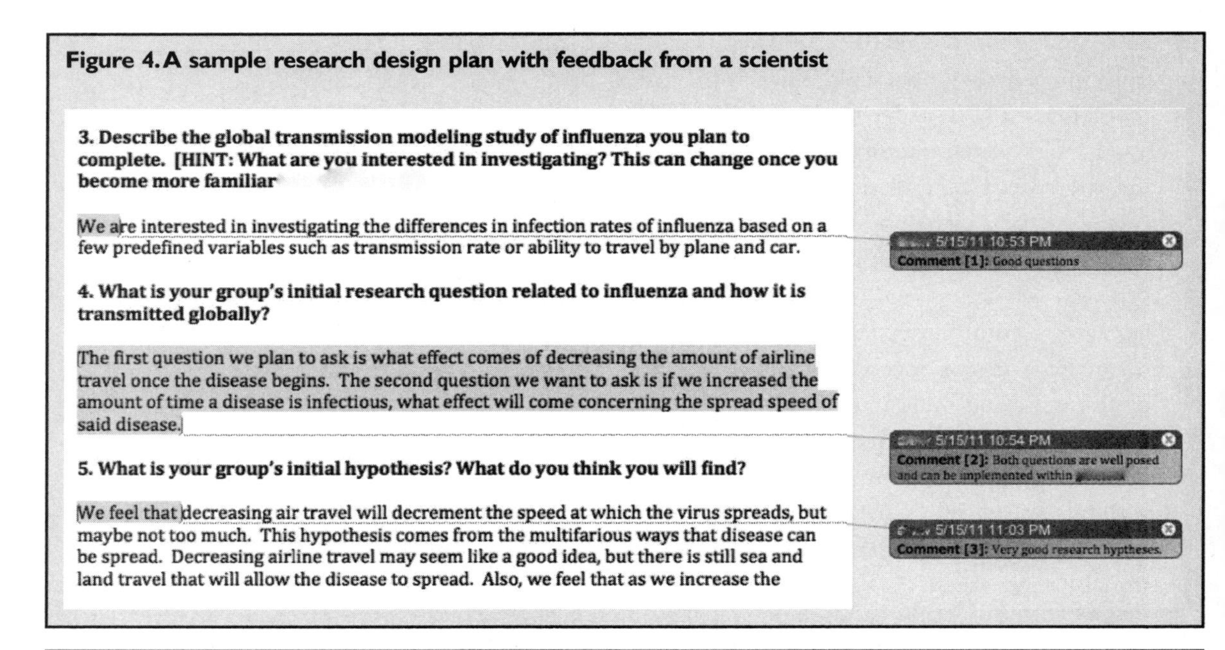

**Figure 4. A sample research design plan with feedback from a scientist**

**3. Describe the global transmission modeling study of influenza you plan to complete. [HINT: What are you interested in investigating? This can change once you become more familiar**

We are interested in investigating the differences in infection rates of influenza based on a few predefined variables such as transmission rate or ability to travel by plane and car.

> 5/15/11 10:53 PM
> **Comment [1]:** Good questions

**4. What is your group's initial research question related to influenza and how it is transmitted globally?**

The first question we plan to ask is what effect comes of decreasing the amount of airline travel once the disease begins. The second question we want to ask is if we increased the amount of time a disease is infectious, what effect will come concerning the spread speed of said disease.

> 5/15/11 10:54 PM
> **Comment [2]:** Both questions are well posed and can be implemented within

**5. What is your group's initial hypothesis? What do you think you will find?**

We feel that decreasing air travel will decrement the speed at which the virus spreads, but maybe not too much. This hypothesis comes from the multifarious ways that disease can be spread. Decreasing airline travel may seem like a good idea, but there is still sea and land travel that will allow the disease to spread. Also, we feel that as we increase the

> 5/15/11 11:03 PM
> **Comment [3]:** Very good research hyptheses.

**Figure 5. An elevator speech summarizing a social network analysis study**

**We work together to use social network analysis to study how a disease spreads through a population. Our research question is, "Can we find certain 'hotspot' people who we can immunize in order to prevent the spread of disease and prevent an epidemic?" Right now, we have found that we can definitely isolate certain people whom we can immunize to shut down the social network system and make it harder for the disease to spread from person to person. These people who have the highest betweenness centrality and who are the most connected to everyone else.**

After receiving this additional feedback on their research, students wrote an original mini-research paper in which they fused aspects of their research design plan with their data analysis. They drafted findings and crafted evidence-based arguments to make claims related to their research questions. These claims were undergirded by their data and analyses of those data. These mini-research papers were peer-reviewed and published online so that others ranging from teachers to peers to parents to others in the community could read about their work.

## Conclusions

We hope this article can open up a discussion with science educators in all areas of the system—from K–12 schools to informal science institutions and afterschool learning environments—about the varied ways to provide opportunities for young people to obtain, evaluate, and communicate information in science and engineering. Substantial acts of reading, writing, and otherwise communicating should be embedded in students' science and engineering investigations. As described in the *Framework,* this supports important cognitive and social learning processes, it helps accomplish the ambitious learning goals outlined in the *Framework,* and it also allows related learning goals to be focused on (e.g., those outlined in the *Common Core State Standards* in mathematics and English language arts—science and technology). For these reasons, it is an ideal time to engage youth in practices related to obtaining, evaluating, and communicating scientific and engineering-related information.

*Philip Bell* is professor of the learning sciences at the University of Washington, Seattle. He served on the team that developed the NRC *Framework*. *Leah Bricker* is assistant professor, science education, at the University of Michigan. *Carrie Tzou* is an assistant professor, science education, at the University of Washington, Bothell; *Tiffany Lee* (is a postdoctoral scholar with the Learning in Informal and Formal Environments (LIFE) Center at the University of Washington, Seattle; and *Katie Van Horne* is a graduate researcher at the University of Washington Institute for Science and Math Education.

## References

Ballinger, C., and M. Carpenter. 2004. The puzzling child: challenging assumptions about participation and meaning in talking science. *Language Arts* 81 (4): 303–311.

Bell, P. 2004. Promoting students' argument construction and collaborative debate in the science classroom. In *Internet environments for science education,* ed. M. C. Linn, E. A. Davis, and P. Bell, pp. 115–143. Mahwah, NJ: Erlbaum.

Bell, P., and M. C. Linn. 2002. Beliefs about science: How does science instruction contribute? In *Personal epistemology: The psychology of beliefs about knowledge and knowing*, ed. B. K. Hofer and P. R. Pintrich, pp. 321–346. Mahwah, NJ: Lawrence Erlbaum.

Calabrese Barton, A. 1998. Reframing "science for all" through the politics of poverty. *Educational Policy* 12 (5): 525–541.

Linn, M. C., and S. Hsi. 2000. *Computers, teachers, peers: Science learning partners.* Mahwah, NJ: Lawrence Erlbaum Associates.

National Research Council (NRC). 2012. *A framework for K–12 science education: Practices, crosscutting concepts, and core ideas.* Washington, DC: National Academies Press.

Tenopir, C., and D. W. King. 2004. *Communication patterns of engineers.* Hoboken, NJ: Wiley.

Tzou, C., and P. Bell. 2010. Micros and Me: Leveraging home and community practices in formal science instruction. In *Learning in the Disciplines: Proceedings of the 9th International Conference of the Learning Sciences (ICLS 2010)—Volume 1, Full Papers,* ed. K. Gomez, L. Lyons, and J. Radinsky, pp. 1,135–1,142. Chicago, IL: International Society of the Learning Sciences.

Warren, B., M. Ogonowski, and S. Pothier. 2003. "Everyday" and "scientific:" Rethinking dichotomies in modes of thinking in science learning. In *Everyday matters in mathematics and science education: Studies of complex classroom events,* ed. A. Nemirovsky, A. Rosebery, J. Solomon, and B. Warren. Mahwah, pp. 119–152. NJ: Erlbaum.

# Making Connections in Math With the *Common Core State Standards*

*By Robert Mayes and Thomas R. Koballa, Jr.*

The vision for science education set forth in *A Framework for K–12 Science Education* (*Framework*; NRC 2012) makes it clear that for today's students to become the scientifically literate citizens of tomorrow their educational experiences must help them become mathematically proficient. "The focus here is on important practices, such as modeling, developing explanations, and engaging in critique and evaluation" (NRC 2012, p. 44). Mathematics is fundamental to modeling and providing evidence-based conclusions. The *Framework* also includes "using mathematics, information and computer technology, and computational thinking" in its list of eight essential practices for K–12 science and mathematics (NRC 2012, p. 49). But what does it mean for students to become mathematically proficient in the context of science? And how can science teachers help students develop that proficiency? This article addresses these questions.

To many Americans, mathematical proficiency means being able to robotically calculate or apply algorithms. Yet, the *Common Core State Standards* (*CCSS*) highlight a very different view. "Mathematically proficient students can apply the mathematics they know to solve problems arising in everyday life, society, and the workplace" (NGAC and CCSSO 2010, p. 7). It's this view of mathematical proficiency that permeates the three dimensions of science presented in the *Framework*.

Inherent in this view is quantitative reasoning, which includes (1) the act of quantification where students identify variables within a context, with attributed units of measure, (2) the use of mathematical concepts in ways that enable description, manipulation, and the generation of claims from quantifiable variables, (3) the use of mathematical models to discover trends and make predictions, and (4) the creation and revision of mathematical representations of phenomena (Mayes, Peterson, and Bonilla 2012).

## Science, engineering, and mathematical practices

To provide a glimpse of how this view of mathematical proficiency will become an important element of the future science education of K–12 students, let's focus on the *Framework's* scientific and engineering practices (SEP) and the mathematical practices (MP) of the *CCSS, Mathetmatics* (*CCSS-M*). (Their alignment is shown in Figure 1 on page 92.)

## Asking and investigating questions

Developing students' ability to ask well-formulated questions is basic to both science and engineering (Practice 1) and mathematics (Practice 1). The *CCSS-M* call for students to be able to determine the meaning of a problem and find entry points to its solution, which requires analyzing givens, constraints, relationships, and goals, with the purpose of making conjectures (i.e., formulating hypotheses) to be tested. Just as science requires formulation and refinement

**Table 1. Alignment between mathematical practices (MP) and scientific and engineering practices (SEP)**

| MP | SEP |
|---|---|
| 1. Making sense of problems and persevering in solving them | 1. Asking questions and defining problems<br>3. Planning and carrying out investigations |
| 2. Reasoning abstractly and quantitatively | 2. Developing and using models<br>3. Planning and carrying out investigations<br>5. Using mathematics and computational thinking |
| 3. Constructing viable arguments and critiquing the reasoning of others | 5. Using mathematics and computational thinking<br>6. Constructing explanations and designing solutions<br>7. Engaging in argument from evidence<br>8. Obtaining, evaluating, and communicating information |
| 4. Model with mathematics | 2. Developing and using models<br>3. Planning and carrying out investigations |
| 5. Using appropriate tools strategically | 2. Developing and using models<br>3. Planning and carrying out investigations<br>4. Analyzing and interpreting data |
| 6. Attending to precision | 3. Planning and carrying out investigations<br>8. Obtaining, evaluating, and communicating information |
| 7. Looking for and making use of structure | 4. Analyzing and interpreting data<br>6. Constructing explanations and designing solutions<br>7. Engaging in argument from evidence |
| 8. Looking for and expressing regularity in repeated reasoning | 5. Using mathematics and computational thinking<br>6. Constructing explanations and designing solutions |

of questions so they can be answered empirically, mathematics attends to questions that may be quantified and then addressed mathematically. Making sense of problems and persevering in solving them (MP 1) calls for the conjectures to be followed by planning a means to reach a solution. Students should consider analogous problems, test special cases, and decompose the problem into simpler cases. This parallels the science focus on designing experimental or observational inquiries (planning and carrying out investigations, SEP 3). This process begins by quantifying the situation being studied through identifying variables and considering how they can be observed, measured, and controlled, as well as considering confounding variables. Students should be engaged in investigations that emerge from their own questions about real-world grand challenges, such as availability and uses of energy resources or biodiversity loss, which are related to their community or region. The interdisciplinary nature of such questions will lead naturally to linkages between science and mathematics.

## Problem solving

Making sense of problems and persevering in solving them (MP 2) calls for students to make sense of quantities and their relationships in problem situations. This typically unfolds in three steps:

1. Students must be able to identify quantities within a scientific context, then represent the situation symbolically. This sets the stage for manipulating the variable quantities using rules of mathematics.
2. Students must continually relate the variables and mathematical representations to the science context so the manipulations they perform move them closer to answering the posed question.
3. Students must move back to the scientific context to provide a data-based solution to the problem.

The act of quantification is essential to creating variables and ensuring that a variable has attributes and measure. It deserves more attention in science classrooms than currently given.

## Models and modeling

Both the *Framework* and *CCSS-M* call for a focus on modeling. Models as discussed in the *Framework* are more broadly construed as diagrams, physical replicas, analogies, computer simulations, and mathematical representations. *CCSS-M* emphasizes abstract mathematical reasoning and quantitative reasoning with the goal of developing an abstract mathematical model such as an equation or function.

Not all models are necessarily quantitative, and quantitative models can take on many different representations beyond that of an equation. Quantitative models can be tables of data, graphs of relationships, statistical displays such as pie graphs, and pictorial science models such as the carbon cycle model shown in Figure 2 (p. 94), adapted from one by the GLOBE Carbon Cycle Project. In addition, science often has embedded variables in models, something outside the experience of students in a mathematics class where typically only two variables are displayed.

Finally, in mathematics, students are too often provided with data or with the equation modeling data without engaging them in collecting data. Data collection is a scientific and engineering practice (planning and carrying out investigations, SEP 3) that is a natural extension of the investigative design process. The design process involves students determining what data are to be gathered, what instruments are needed to measure data, how much data are needed to address reliability and precision concerns, and what experimental procedures to follow. The processes of designing investigations and collecting data have the potential to engage students in all the MPs, 1 through 6.

**Figure 2. Carbon cycle model**

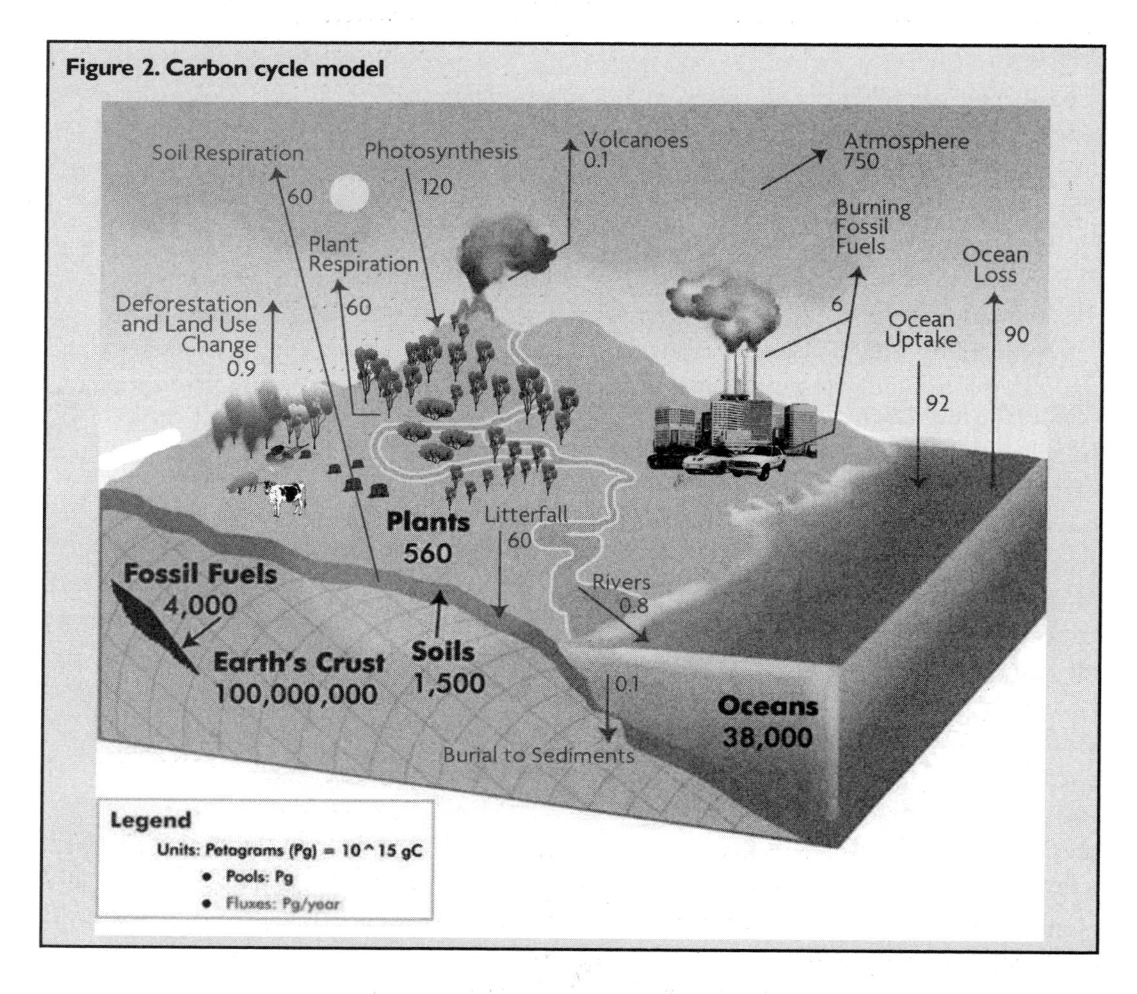


## Argumentation

Constructing viable arguments and critiquing the reasoning of others (MP 3) is analogous to the scientific and engineering practice of engaging in argument from evidence (SEP 7). Both emphasize justifying claims in an argument by grounding them in evidence (e.g., mathematical or scientific theories) that is accepted by the scientific or mathematical community. Students should critique their own arguments, identifying weaknesses and flaws in logic, revise their arguments, and submit them to peers for review. The ultimate form of argument in mathematics is the abstract proof, but truly understanding a proof requires experimentation, just as in science. Data analysis is fundamental to argumentation in science and engineering, but it often requires quantitative analysis of data within a context. Both science and mathematics value the ability to compare the effectiveness of two plausible arguments and distinguish correct logical reasoning from flawed reasoning.

In addition to constructing an argument, communicating it to others is paramount (SEP 8). The mathematical practices of constructing viable arguments and critiquing the reasoning of others (MP 3) and attending to precision (MP 6) have elements of communication, the ability to communicate precisely and clearly a mathematical argument to peers and experts. MP 6 highlights the special difficulties in communicating mathematics due to the intensively symbolic nature of the subject as well as the density of language in mathematical texts. SEP 8 focuses on related communication issues, specifically attending to the difficulty students have with reading scientific and engineering texts and primary literature, as well as with scientific writing. The difficulty of communicating in science and mathematics is compounded by the fact that scientific ideas are often represented quantitatively as tables, graphs, charts, equations, and symbols.

## Mathematical tools use

The selection of appropriate mathematical tools for certain tasks is the focus of MP 5. In relation to science, the toolbox includes selection of the appropriate type of model (SEP 2). Students must understand the limitations and precision of the selected model type, for example, using a graphic representation rather than an equation. It also means the ability to select the appropriate mathematical algorithm as a tool to analyze data within a science context. It includes the type of instrument used, such as selecting a protractor, calculator with remote light sensor, spreadsheet, computer algebra system, statistical package, dynamic geometry software, or computer simulation. In addition, science and engineering have an extensive set of instruments to select from to measure quantities (planning and carrying out investigations, SEP 3), which raises concerns of precision, accuracy, and error. Once data are collected, they need to be analyzed and interpreted (SEP 4). Mathematics is essential for expressing relationships in the data. Students of science tell the story of data using descriptive statistics, test hypotheses using statistical analysis, and explore causal and correlational relationships.

The toolbox for science and engineering is ever expanding, with the advent of two new paradigms. The paradigms of empirical methods (applied or experimental science) and theory (theoretical science) were until recently considered the two legs of science. But over the last 20 years, due to increasing computing capabilities, two new paradigms have arisen: computational science (scientific computing) and data-intensive science (data-centric science) (Hey, Tansley, and Tolle 2009).

Computational science is embedded in mathematics, science and engineering, and the humanities; it complements the empirical methods and theory paradigms but does not replace them. The goal of scientific computing is to improve the understanding of physical phenomena. Scientific computing focuses on simulations and modeling to provide both qualitative and quantitative insights into complex systems and phenomena that would be too expensive, dangerous, or even impossible to study by direct experimentation or theoretical methods (Turner et al. 2011). The explosion of data in the 21st century led to the invention of data-intensive science as a fourth paradigm, which focuses on compressed sensing (effective use of large data sets), curation (data storage issues), analysis and modeling (mining the data), and visualization (effective human-computer interface). SEP 5 highlights science and engineering education issues related to these two new paradigms.

## Mathematical structure

The ability to look for and make use of structure (MP 7) and look for and express regularity in repeated reasoning (MP 8) focus on abstract mathematical argumentation. For example, a student who can see the structure of the distributive property $a(b+c) = ab + ac$ in the expression $(x+y)(b+c) = (x+y)b + (x+y)c$ does not need to memorize rules for multiplying binomial expressions. Use of structure and repeated reasoning most closely align with constructing explanations and designing solutions (SEP 6). We consider these to be the more theoretical aspects of the mathematics and science processes, while the other processes are more experimental in nature. Students need to engage with standard scientific explanations of the world that link science theory with specific observations.

While we find a lot of commonality between the practices put forth in the *Framework* and the *CCSS-M*, we share concerns similar to those discussed in a review of the Summer 2012 draft of the *Next Generation Science Standards* (*NGSS*) conducted by the Fordham Institute (Gross et al. 2012). In brief, the Fordham Institute reviewers revealed the need for mathematics content specificity in the *NGSS* and for vigilance in the alignment of the *NGSS* and the *CCSS-M* to achieve the desired dovetailing of science and mathematics learning across the grade levels.

We believe it is only through careful attention to the specific science, engineering, and mathematics concepts to be learned and the alignment of them across the grade levels that the vision for science and engineering teaching and learning presented in the *Framework* can be realized. In our examples that follow, we attempt to illustrate how this attention and alignment might be enacted.

## Examples of *Framework* and *CCSS-M* alignment

The core Earth and Space Science idea of Earth and Human Activity provides a good context for showcasing grand challenges that students can explore in their own community. Concepts central to this core idea include natural resources, natural hazards, human impact on Earth systems, and global climate change. Change is a core quantitative concept, so we chose global climate change as our focus concept. Following the lead of the *Framework*, we discuss science tasks that could be accomplished by the end of grades 2, 5, 8, and 12. The endpoints for these grades described in the *CCSS-M* and in the *Framework* for Global Climate Change are presented in Figure 3.

### Grade 2

Climate is not a grade 2 concept. However, prerequisite understandings are developed through the study of weather. Weather tasks for second graders that address both sets of understandings might involve students observing television weather reports followed by drawing pictures of and describing things they believe make up the weather. These experiences will enable students to construct their own definitions of weather and list variables that make up weather, such as rain, sunshine, and wind.

**Figure 3. Examples of *Framework* and *CCSS-M* alignment for global climate change**

| Grade Level | *Framework for K–12 Science Education* | *Common Core State Standards, Mathematics* |
|---|---|---|
| **Grade 2** | By the end of grade 2 students should know that "Weather is the combination of sunlight, wind, snow or rain, and temperature in a particular region at a particular time. People measure these conditions to describe and record weather and to notice patterns over time" (NRC 2012, p. 188). | The *CCSS-M* have second graders solving problems involving addition and subtraction within 100, understanding place value up to 1,000, recognizing the need for standard units of measure of length, representing and interpreting data, and reasoning with basic shapes and their attributes. |
| **Grade 5** | By the end of fifth grade the expectation for global climate change is, "If Earth's global mean temperature continues to rise, the lives of humans and organisms will be affected in many different ways" (NRC 2012, p. 98). | The *CCSS-M* has fifth graders writing and interpreting numerical expressions, analyzing patterns and relationships, performing operations with multi-digit whole numbers and decimals to hundredths, using equivalent fractions to add and subtract fractions, multiplying and dividing fractions, converting measurement units within a given measurement system, measuring volume, representing and interpreting data, graphing points on the coordinate plane to solve real-world problems, and classifying two-dimensional figures into categories based on their properties. |
| **Grade 8** | The end of eighth grade expectation for climate change is to understand that human activities, such as carbon dioxide release from burning fuels, are major factors in global warming. Reducing the level of climate change requires an understanding of climate science, engineering capabilities, and human behavior (NRC 2012, p. 198). | The *CCSS-M* eighth grade standards include awareness of numbers beyond the rational numbers, work with radicals and integer exponents, proportional relationships, ability to analyze and solve linear equations and systems of linear equations, use linear functions to model relationships between quantities, understand congruence and similarity, the Pythagorean Theorem, solve real-world problems involving volume of cylinders, cones, and spheres, and use statistics to investigate patterns of association in bivariate data. |
| **Grade 12** | By the end of high school students should understand that climate change is slow and difficult to recognize without studying long-term trends, such as studying past climate patterns. Computer simulations are providing a new lens for researching climate change, revealing important discoveries about how the ocean, the atmosphere, and the biosphere interact and are modified in response to human activity (NRC 2012, p. 198). | The *CCSS-M* high school standards are by conceptual categories not grade level. The conceptual categories of Number and Quantity, Algebra, Functions, Modeling, Geometry, and Statistics and Probability specify the mathematics that all students should study in order to be college and career ready. Functions are expanded to include quadratic, exponential, and trigonometric functions, broadening the potential models for science. |

**Figure 4. National temperature trends (Grade 5)**

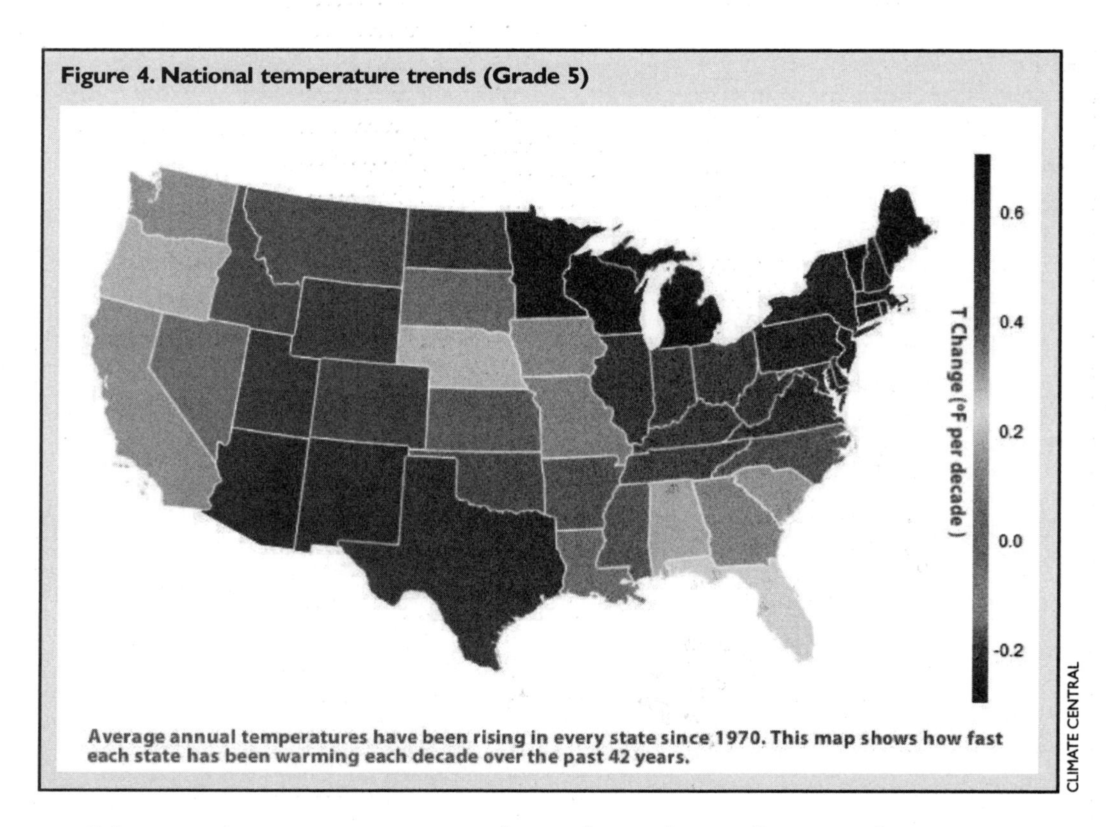


Average annual temperatures have been rising in every state since 1970. This map shows how fast each state has been warming each decade over the past 42 years.

Subsequent learning experiences might involve students collecting and measuring rain to the nearest centimeter for each month of the school year for their community or being given these data. Then, students could be asked to draw pictures representing rain by month; this may be a bar graph or a dot chart using M&M candies. Using visual data displays, student could answer questions about specific weather variables: Which month was the wettest? the driest? Conclude by having students link their findings to the context of the local environment through such questions as these: What do you think happened to plants in the months with low rainfall? What other weather conditions interact with the amount of rain to affect plant life?

## Grade 5

Climate Change tasks for fifth grade may have students considering data on state, national, and international annual temperature changes. For example, students could be asked to examine Climate Central's national map on temperature change (Figure 4; also see "On the web"). Questions to prompt their interactions with the map could include "What percentage of states has warmed more than 0.2 degrees each decade over the past 40 years? How much has the state you lived in warmed?"

Further investigations might have students examining data for the state in which they live. For example, students could be directed to one of the red points on the graph representing Georgia (Figure 5) and asked to interpret what it means. What does the general trend of the scatter plot of points indicate? Further using the information presented in the figure, students might be asked to measure the temperature each day for a week to the nearest 0.1 degree. What can you say about natural flux in daily temperatures and how it relates to the annual average temperature? If the temperature continues to increase at the current rate, what will the average temperature be in 20 years? What potential impact does this warming trend have in your state? Sample responses could be decreased biodiversity due to extinctions, agricultural economic impact, and increased heat-related problems for the football team.

**Figure 5. Georgia temperature trends (Grade 8)**

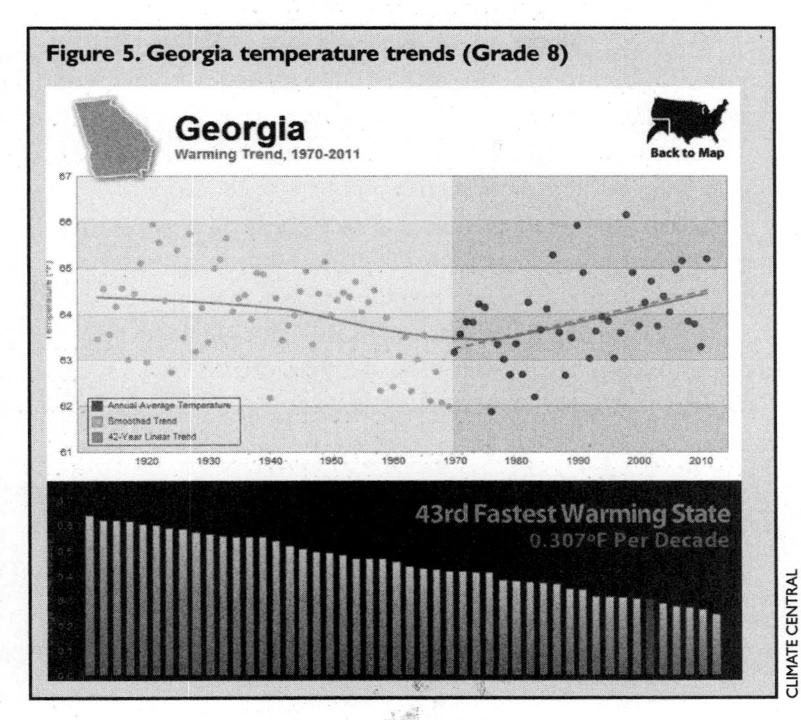


CLIMATE CENTRAL

## Grade 8

Climate Change tasks for eighth-grade students could be initiated by extending the discussion of the Georgia warming data. Provide students with the data for average annual temperature per year for the state in a table, then have them plot the data and construct a scatter plot like the one in Figure 5. The plot could then be used to address questions such as these: "What is the trend of the data in this scatter plot? Is it decreasing or increasing? Estimate a line of best fit for the data that represents the trend." Discussing with students what is needed to determine a line, slope, and a point may help them accomplish this. A potential point for the line is the center point of the data set, which students can calculate as the ordered pair with $x$-intercept (the average of the first coordinate values of the data points in the set) and the $y$-intercept (the average of the second coordinate values).

Once students have determined the center point of the data set, ask them to place a ruler on the center point and vary the slope by rotating the ruler about the center point to best represent the trend of the data. Then, have them write out the equation of the line and use the linear model to predict temperatures for future years. Conclude by helping students relate this back to the science context, using such questions as "What variables can we control to reduce or stabilize the temperature trend?" Among the possible variables is carbon dioxide, which, if controlled or reduced, would reduce greenhouse gases in the atmosphere and may impact climate change.

### Grade 12

Climate Change tasks for grade 12 students may involve revisiting the scatter plot of state temperature data. But, this time ask students to provide a power function model or exponential model for the data. Rich discussions of which function is the best model for the data would engage students in exploring error and best-fit concepts.

Carbon dioxide as a mitigating factor in global climate change can be explored in more depth. For example, Figure 6 provides data on historic trends in atmospheric carbon dioxide. Ask students to quantitatively interpret the trends in the graph as naturally occurring cycles. The claim has been made that today the Earth is experiencing just a phase in a natural cycle of carbon dioxide change. Students could be challenged to interpret the data for evidence that supports this claim. Questions that could serve to guide students' work include "How were the data collected? Are the data reliable? What are likely causes of the fluxes in atmospheric carbon dioxide?"

**Figure 6. Atmospheric carbon dioxide flux (Grade 12)**

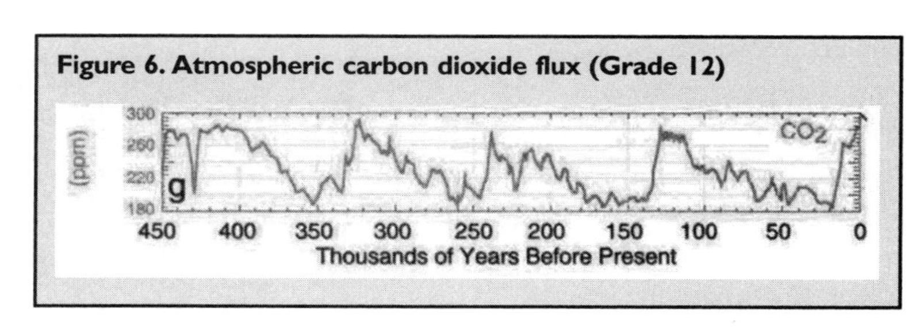


## Looking forward

There is much work yet be done to specify the science, engineering, and mathematics concepts to be learned by K–12 students and align them across the grade levels, but there is great promise in this work. The approaching release of the *NGSS* on the heels of the *CCSS-M* facilitates this work, making alignment and compatible pacing of expected learning outcomes across the grade levels possible. Both the *Framework* and the *CCSS-M* emphasize student construction of conceptual understandings and the development of real-world practices. Indeed, mathematical proficiency in the context of science highlights the application of mathematics to solve real-world problems in everyday life, society, and the workplace.

Through the *Framework* we see glimpses of curricula that are problem based and community focused to increase student engagement and that take interdisciplinary approaches to the grand challenges of today and tomorrow. These curricula embrace mathematical proficiency as a desirable outcome of science and engineering education. Students must build mathematical proficiency as they develop understandings and skills in science and engineering that will enable them to become the scientifically literate citizens of tomorrow.

Helping students develop the mathematical proficiency described in the *Framework* will be challenging. However, you can take steps to enhance the odds of your success. We recommend that science teachers read the *CCSS-M* and upon their release, the *NGSS*, focusing on the guidance most pertinent to the grade level or high school courses you teach. Look for points of alignment and compatibility of pacing and determine what will work for your school context and students. The information presented on grade endpoints in *CCSS-M* and the *Framework* may serve as an advance organizer for this exploration of the standards documents. We also encourage meeting regularly with mathematics teachers to discuss expectations for student learning and work collaboratively to build lessons and units.

As you progress with this work, recognize that while standards offer guidance, it is teachers, through planning and instruction, who enact the vision for student success set forth in the standards. Finally, we recommend that science teachers strengthen their own understandings of the mathematics germane to the science they teach. This can be done by establishing professional learning communities (PLC) that are interdisciplinary, including both science and mathematics teachers. The PLC can review the *NGSS* and *CCSS-M* standards together, selecting real-world grand challenges to engage students in cross-discipline, problem-based episodes. School administrations can support the process by providing common planning time and, when possible, team teaching of science, technology, engineering, and math (STEM) courses. In addition, the PLC can reach out to regional higher education institutions and STEM research centers to seek mentoring from scientists and mathematicians on STEM content.

*Robert Mayes* directs the Institute for Interdisciplinary STEM Education; *Thomas Koballa* is dean of the College of Education, both at Georgia Southern University in Statesboro, Georgia.

## On the web

Climate Central national map: *www.climatecentral.org/news/the-heat-is-on*

## References

Gross, P. R., L. S. Lerner, J. Lynch, M. Schwartz, R. Schwartz, and W. S. Wilson. 2012. *Commentary and feedback on draft I of the* Next Generation Science Standards. Washington, DC: Fordham Institute.

Hey, T., S. Tansley, and K. Tolle, eds. 2009. *The fourth paradigm: Data intensive scientific discovery.* Redmond, WA: Microsoft Research.

Mayes, R., F. Peterson, and R. Bonilla. 2012. Quantitative reasoning: Current state of understanding. In *WISDOMe: Quantitative reasoning and mathematical modeling: A driver for STEM integrated education and teaching in context,* ed. R. Mayes and L. Hatfield, pp. 7–38. Laramie, WY: University of Wyoming.

National Governors Association Center for Best Practices and Council of Chief State School Officers (NGAC and CCSSO). 2010. *Common core state standards (mathematics).* Washington, DC: NGAC and CCSSO.

National Research Council (NRC). 2012. *A framework for K–12 science education: Practices, crosscutting concepts, and core ideas.* Washington, DC: National Academies Press.

Turner, P., et al. 2011. Undergraduate computational science and engineering education. *Society for Industrial and Applied Mathematics Review* (SIAM Rev.) 53 (3): 561–574.

# The *Next Generation Science Standards* and the Life Sciences

***By Rodger W. Bybee***

Publication of the *Next Generation Science Standards* (*NGSS*) will be just short of two decades since publication of the *National Science Education Standards* (NRC 1996). In that time, biology and science education communities have advanced, and the new standards will reflect that progress (NRC 1999, 2007, 2009; Kress and Barrett 2001).
Just as earlier standards influenced state-level standards, assessments, and science teachers at all levels, so too will the *NGSS*.

Using the life sciences, this article first reviews essential features of the NRC *Framework for K–12 Science Education* (*Framework*; NRC 2012) that provided a foundation for the new standards. Second, the article describes the important features of life science standards for elementary, middle, and high school levels. Finally, I discuss several implications of the new standards. This article extends other discussions of biology and the *NGSS* (Bybee 2011, 2012) and other publications for science teachers (Willard et al. 2012).

Core ideas for the life sciences consist of the following: From Molecules to Organisms, Ecosystems, Heredity, and Biological Evolution. The following sections describe the core and component ideas for K–12 life sciences (NRC 2012) in greater detail.

## Disciplinary core ideas for the life sciences

### Core Idea 1: From Molecules to Organisms: Structures and Processes

This core idea addresses the characteristic structures of organisms. Individual organisms also accomplish specific functions to support life, growth, behavior, and reproduction. This core idea centers on the unifying principle that cells are the basic unit of life. This core idea includes the following component ideas:

*Structure and function:* Beginning with cells as the basic structural units of life, organisms present a hierarchy of structural systems and subsystems that perform specialized functions. A central problem of biology is to develop explanations for functions based on structures and the reciprocal—to explain the complementarity of structures and functions among an organism's systems and subsystems.

*Growth and development of organisms:* As organisms grow and develop their anatomy and morphology (structures), processes from the molecular to cellular to organism-level, as well as behaviors, change in predictable ways. Central to understanding growth and development of organisms are the concepts of cell division and gene expression.

*Organization for matter and energy flow in organisms:* Organisms require matter and energy in order to live and grow. In most cases the energy needed by organisms is derived from the Sun through photosynthesis. As a result of chemical changes, energy is transferred from one system of interacting molecules to another and across different organizational levels from cells to ecosystems.

*Information processing:* Organisms have mechanisms to detect, process, and use information about the environment. That information contributes to an organism's survival, growth, and reproduction.

### Core Idea 2: Ecosystems: Interactions, Energy, and Dynamics

This core idea includes organisms' interactions with each other and their physical environment. Biologists develop explanations for how organisms obtain resources, how they change their environment, how changing environmental factors affect organisms and ecosystems, how social interactions and group behavior play out within and between species, and how these factors all combine to determine ecosystem functioning. This core idea includes the following component ideas.

*Interdependent relationships in ecosystems:* An ecosystem includes both biological communities (biotic) and physical (abiotic) components of the environment. Ecosystems continually change due to the interdependence of biotic and the abiotic elements of the environment. As organisms seek matter and energy to sustain life, the interactions may be represented as food webs.

*Cycles of matter and energy transfer in ecosystems:* Interactions among organisms and the physical environment influence the cycling of matter and flow of energy in ecosystems. Plants require light energy for photosynthesis—a chemical reaction that produces plant matter from air and water. As animals meet their need for food, the chemical elements that make up organisms are combined and recombined as those chemical elements pass through food webs. The cycling of matter and flow of energy through ecosystems conserve matter and energy through the many changes.

*Ecosystem dynamics, functioning, and resilience:* Dynamics of ecosystems result from changes in populations of organisms through time and changes in physical environments. The dynamics of ecosystems result in shifts such as changes in the diversity and numbers of organisms, the survival or extinction of species, the migration of species, and the evolution of new species. Changes in ecosystems can result from natural processes and human activity. The resilience of an ecosystem is a function of greater or lesser biodiversity.

*Social interactions and group behavior:* Organisms ranging from unicellular slime molds to humans demonstrate group behavior. Group behavior can be explained by its survival value for individuals.

### Core Idea 3: Heredity: Inheritance and Variation of Traits

This core idea focuses on the flow of genetic information between generations. It explains the mechanisms of genetic inheritance and describes the environmental and genetic causes of gene mutation and the alteration of gene expression. This core idea includes the following component ideas.

*Inheritance of traits:* Heredity refers to the processes by which characteristics of a species are passed from one generation to the next. Heredity explains why offspring look like, but are not identical to, parents.

Chromosomes carry the genetic information for a species' characteristics. Each chromosome consists of a single DNA molecule, and each gene is a particular segment of DNA. DNA molecules consist of four building blocks called nucleotides that form a linked sequence. The specific sequence of nucleotides constitutes a gene's information. Through cellular processes, that genetic information forms proteins, which build an organism's characteristics.

*Variation of traits:* Genetic and environmental factors produce variations of traits within a species population. Variation in traits can influence the development, appearance, behavior, and ability of organisms to produce offspring. The distribution of variations of traits in a population is an essential factor in biological evolution.

## Core Idea 4: Biological Evolution: Unity and Diversity

This core idea uses "changes in the traits of populations of organisms over time" to explain species' unity and diversity. Biological evolution is supported by extensive scientific evidence ranging from the fossil record to genetic relationships among species. This core idea includes the following component ideas:

*Evidence of common ancestry and diversity:* Biological evolution results from changing environmental factors and the subsequent selection from among genetic variations in a population that across generations changes the distribution of those characteristics in that population.

Common ancestry and diversity are supported by multiple lines of empirical evidence including the fossil record, comparative anatomy and embryology, similarities of cellular processes and structures, and comparisons of DNA sequences between species. Recent advances in molecular biology have provided new empirical evidence supporting prior explanations for changes in the fossil record and links between living and extinct species.

*Natural selection:* As environments change, organisms with variations of some traits may be more likely than others to survive and reproduce. Genetic variation in a species makes this process of natural selection possible. In time, natural selection results in changes in the distribution of certain traits. That is, selection leads to an increase of organisms in a population with certain inherited traits and a decrease in other traits.

*Adaptation:* Natural selection is the mechanism by which species adapt to changes in resources or the physical limits and biological challenges an environment imposes. In the course of many generations adaptation can result in the formation of new species. If a population cannot adapt due to a lack of traits that contribute to survival and reproduction, the species may become extinct.

*Biodiversity and humans:* Biodiversity is the multiplicity of genes, species, and ecosystems. It provides humans with renewable resources and benefits such as ecosystem services. Biological resources must be used within sustainable limits or there will be detrimental consequences such as ecosystem degradation, species extinction, and reduction of ecosystem services.

The four core ideas for the life sciences have a long history and solid foundation as the basis for the life sciences in school programs (Hurd 1961; Bybee and Bloom 2008; BSCS 1993). These core ideas extend and elaborate those established K–12 science education standards: *National Science Education Standards* (NRC 1996) and *Benchmarks for Science Literacy* (AAAS 1993). The ideas also incorporate the *Science College Board Standards for College Success* (College Board 2009), and the ideas are consistent with frameworks for national and international assessments.

## From the *Framework* to standards

The NRC *Framework* provided guidance for developing standards through 13 recommendations designed to ensure fidelity to the *Framework* and serve as direction for the development

**Figure 1. Essentials of *A Framework for K–12 Science Education***

*A Framework for K–12 Science Education: Practices, Crosscutting Concepts, and Core Ideas* (NRC 2012) presents fundamental concepts and practices for the new standards and implied changes in K–12 science programs. The *Framework* describes three essential dimensions: science and engineering practices, crosscutting concepts, and core ideas in science disciplines. In this article, the core disciplinary ideas are from the life sciences.

The scientific and engineering practices were discussed earlier in this book (see the "Scientific and Engineering Practices in K-12 Classrooms") and are summarized below.

**Practices for K–12 science curriculum**

1. Asking questions (for science) and defining problems (for engineering)
2. Developing and using models
3. Planning and carrying out investigations
4. Analyzing and interpreting data
5. Using mathematics and computational thinking
6. Constructing explanations (for science) and designing solutions (for engineering)
7. Engaging in argument from evidence
8. Obtaining, evaluating, and communicating information

The second dimension described in the NRC *Framework* is crosscutting concepts. These too have been discussed in the earlier chapter by Duschl and are summarized here.

**Crosscutting concepts for K–12 science education**

1. *Patterns.* Observed patterns in nature guide organization and classification and prompt questions about relationships and causes underlying the patterns.
2. *Cause and effect: Mechanism and explanation.* Events have causes, sometimes simple, sometimes multifaceted. Deciphering causal relationships and the mechanisms by which they are mediated is a major activity of science.
3. *Scale, proportion, and quantity.* In considering phenomena, it is critical to recognize what is relevant at different sizes, times, and energy scales and to recognize proportional relationships between different quantities as scales change.
4. *Systems and system models.* Delimiting and defining the system under study and making a model of it are tools for developing understanding used throughout science and engineering.
5. *Energy and matter: Flows, cycles, and conservation.* Tracking energy and matter flows, into, out of, and within systems, helps one understand a system's behavior.
6. *Structure and function.* The way an object is shaped or structured determines many of its properties and functions.
7. *Stability and change.* For both designed and natural systems, conditions of stability and what controls rates of change are critical elements to understand.

of standards. For this discussion the following summarizes the National Research Council recommendations for standards development.

The standards should

- Set rigorous goals for all students.
- Be scientifically accurate.
- Be limited in number.
- Emphasize all three dimensions.
- Include performance expectations that integrate the three dimensions.
- Be informed by research on learning and teaching.
- Meet the diverse needs of students and states.
- Have potential for a coherent progression across grades and within grades.
- Be explicit about resources, time, and teacher expertise.
- Align with other K–12 subjects, especially the *Common Core State Standards*.
- Take into account diversity and equity (NRC 2012).

Given the criteria and constraints for developing life science standards, a working group of biology teachers and other educators developed standards for the four unifying concepts and component ideas.* Figures 2 through 4 are example standards for elementary, middle, and high school life sciences, respectively.

The architecture seen in Figures 2 through 4 requires clarification. The titles "From Molecules to Organisms: Structures and Processes," "Biological Evolution: Unity and Diversity," and "Biological Evolution: Unity and Diversity," represent one standard each for elementary, middle, and high school life sciences. The standards include the performance expectations in the top portion, identified as "3-LS1.A," "MS-LS4.F," and "HS-LS4.B" and "HS-LS4.D" in the three figures, respectively. The performance expectations are formed by combining a science and engineering practice, disciplinary core idea, and crosscutting concept.

Immediately beneath the performance expectations, you see the foundation box consisting of three sections, one each for science and engineering practices, disciplinary core ideas, and crosscutting concepts. These three columns present content from the *Framework* and serve as a reference for the performance expectations in the standard. You should note the relationship between "3-LS1.A" "MS-LS4.F," and "HS-LS4.B" and "HS-LS4.D" before the performance expectations and at the end of statements in the foundation box. Descriptions in the foundation box answer the questions

- What are the essential knowledge and abilities of the performance expectations?
- What are the specific details of the practices, disciplinary core ideas, and crosscutting concepts that students should know and be able to do?
- What should be emphasized in the science curriculum and classroom instruction?

The performance expectations are learning outcomes, not instructional activities, and they are the basis for assessments. One should note that along with content in the foundation box, they may be the point of departure for backward design of curriculum instruction (Wiggins and McTighe 2005).

The three examples displayed in Figures 1–3 (pp. 106, 108, and 109) serve another purpose in this discussion. That purpose is to show a learning progression from elementary school to high school for biological evolution. Although elementary students are not expected to learn the mechanisms of natural selection, they learn about heredity and the variation of traits—concepts fundamental to biological evolution described in greater detail in the middle and high school life science standards.

Here, I note that other standards, for example about interdependent relations in ecosystems, also contribute to an elementary student's conceptual foundations for biological evolution.

## From standards to curriculum and instruction

From the late 1980s to the early 2000s, teachers of K–12 science and the larger science education community have witnessed an era of standards-based reform. Basically, the idea is to develop clear, comprehensive, and challenging goals for student learning. Review, for example, the aforementioned guidelines for developing the *NGSS*. Beyond learning goals, the implicit assumption is that standards would result in greater alignment among other components of the

**Figure 2. An example of a standard for elementary school life sciences with supporting content from the foundation box and connection box**

**3-LS1 From Molecules to Organisms: Structures and Processes**

Students who demonstrate understanding can:

**3-LS1-a. Construct explanations from evidence that life cycles of plants and animals have similar features and predictable patterns.** [Clarification Statement: Changes organisms go through during their life form a pattern and can be used to predict what might happen next in a different organism. Reproduction is addressed as just one part of the process of birth, growth, development, reproduction, and death.] [Assessment Boundary: Plant reproduction is limited to flowering plants. Evidence should be provided.]

The performance expectations above were developed using the following elements from the NRC document *A Framework for K-12 Science Education*:

| Science and Engineering Practices | Disciplinary Core Ideas | Crosscutting Concepts |
|---|---|---|
| **Constructing Explanations and Designing Solutions**<br>Constructing explanations and designing solutions in 3–5 builds on prior experiences in K–2 and progresses to the use of evidence in constructing multiple explanations and designing multiple solutions.<br>• Use evidence (e.g., measurements, observations, patterns) to construct a scientific explanation or design a solution to a problem. (3-LS1-a)<br>- - - - - - - - - - - - - - - - -<br>**Connections to Nature of Science**<br>**Scientific Knowledge is Based on Empirical Evidence**<br>• Science findings are based on recognizing patterns. (3-LS1-a) | **LS1.B: Growth and Development of Organisms**<br>• Reproduction is essential to the continued existence of every kind of organism. Plants and animals have unique and diverse life cycles that include being born (sprouting in plants), growing, developing into adults, reproducing, and eventually dying. (3-LS1-a) | **Patterns**<br>• Cyclic patterns of change related to time can be used to make predictions. (3-LS1-a) |

Connections to other DCIs in this grade-level: will be added in future version.

Articulation of DCIs across grade-levels: will be added in future version.

Common Core State Standards Connections: [Note: these connections will be made more explicit and complete in future draft releases]

ELA –

**RI.3.10** By the end of the year, read and comprehend informational texts, including history/social studies, science, and technical texts, at the high end of the grades 2–3 text complexity band independently and proficiently. (3-LS1-a)

**SL.4.4** Report on a topic or text, tell a story, or recount an experience in an organized manner, using appropriate facts and relevant, descriptive details to support main ideas or themes; speak clearly at an understandable pace. (3-LS1-a)

Mathematics –

**MP.3** Construct viable arguments and critique the reasoning of others. (3-LS1-a)

**MP.7** Look for and make use of structure. (3-LS1-a)

educational system—curriculum, instruction, assessments, and the professional development of teachers.

In 2001, Elementary and Secondary Education Act (ESEA) legislation—No Child Left Behind (NCLB)—established assessment as an emphasis in the educational system. This shift in emphasis has significantly influenced the systems' components. Assessment has been a primary concern of educators, and curriculum and instruction have been secondary, at best. This shift to NCLB and priorities of English language arts and math has had the unintentional consequence of reducing or eliminating science in elementary schools. I believe we have gone directly from standards to assessments without addressing curriculum and instruction as the teaching and learning connection.

Relative to the *NGSS*, I am particularly concerned about questions science teachers frequently ask: Where are the curriculum materials that will help me implement the standards in my classroom? And will assessments change? These are both critical questions. There are several initiatives relative to assessment or *NGSS*, but few discussions of new instructional materials.

I cannot emphasize enough the need for clear and coherent curriculum and instruction that connects the *NGSS* and assessments. Curriculum materials will be the missing link if they are not developed and implemented. The absence of a curriculum based on the new standards will be a major failure in this era of standards-based reform and assessment-dominated results.

**Figure 3. An example of a standard for middle school life sciences with supporting content from the foundation box and connection box**

**MS-LS4 Biological Evolution: Unity and Diversity**

Students who demonstrate understanding can:

**MS-LS4-f. Use mathematical models to support the explanation of how natural selection over many generations results in changes within species in response to environmental conditions that tend to increase or decrease specific traits in a population.** [Clarification Statement: Emphasis is on using mathematical models to explain trends based on data for changes in populations over time.] [Assessment Boundary: The assessment should provide evidence of students' abilities to explain trends in data for the number of individuals with specific traits changing over time.]

The performance expectations above were developed using the following elements from the NRC document *A Framework for K-12 Science Education*:

| Science and Engineering Practices | Disciplinary Core Ideas | Crosscutting Concepts |
|---|---|---|
| **Developing and Using Models**<br>Modeling in 6–8 builds on K–5 and progresses to developing, using, and revising models to support explanations, describe, test, and predict more abstract phenomena and design systems.<br>• Develop models to describe unobservable mechanisms. (MS-LS4-f)<br><br>**Using Mathematics and Computational Thinking**<br>Mathematical and computational thinking at the 6–8 level builds on K–5 and progresses to identifying patterns in large data sets and using mathematical concepts to support explanations and arguments.<br>• Apply concepts of ratio, rate, percent, basic operations, and simple algebra to scientific and engineering questions and problems. (MS-LS4-f) | **LS4.B: Natural Selection**<br>• Genetic variations among individuals in a population give some individuals an advantage in surviving and reproducing in their environment. This is known as natural selection. It leads to the predominance of certain traits in a population, and the suppression of others. (MS-LS4-e), (MS-LS4-f)<br><br>**LS4.C: Adaptation**<br>• Adaptation by natural selection acting over generations is one important process by which species change over time in response to changes in environmental conditions. (MS-LS4-f), (MS-LS4-h)<br>• Traits that support successful survival and reproduction in the new environment become more common; those that do not become less common. Thus, the distribution of traits in a population changes. (MS-LS4-f) | **Cause and Effect**<br>• Cause and effect relationships may be used to predict phenomena in natural or designed systems. Phenomena may have more than one cause, and some cause and effect relationships in systems can only be described using probability. (MS-LS4-b), (MS-LS4-e),(MS-LS4-f) |

Connections to other DCIs in this grade-level: will be added in future version.

Articulation of DCIs across grade-levels: will be added in future version.

Common Core State Standards Connections:

ELA /Literacy–

**SL.7.5** Include multimedia components and visual displays in presentations to clarify claims and findings and emphasize salient points. (MS-LS4-f)

Mathematics –

**MP.4** Model with mathematics. (MS-LS4-f)

**5.OA** Analyze patterns and relationships. (MS-LS4-f),(MS-LS4-d)

**6.EE** Represent and analyze quantitative relationships between dependent and independent variable. (MS-LS4-f)

**Figure 4. An example of a standard for high school life sciences with supporting content from the foundation box and connection box**

**HS-LS4 Biological Evolution: Unity and Diversity**

Students who demonstrate understanding can:

**HS-LS4-b. Use a model to support the explanation that the process of natural selection is the result of four factors: (1) the potential for a species to increase in number, (2) the heritable genetic variation of individuals in a species due to mutation and sexual reproduction, (3) competition for limited resources, and (4) the proliferation of those organisms that are better able to survive and reproduce in the environment.** [Clarification Statement: Emphasis is on the interrelationship of the four factors that result in natural selection. Mathematical models and simulations of changes in distribution of traits in a population at different times may be used.] [Assessment Boundary: Assessment should provide evidence of students' abilities to explain natural selection in terms of the number of organisms, behaviors, morphology, or physiology factors having a direct effect on survival and reproduction as well as ability to compete for limited resources. Mathematical models may be used to communicate the explanation.]

The performance expectations above were developed using the following elements from the NRC document *A Framework for K-12 Science Education*:

| Science and Engineering Practices | Disciplinary Core Ideas | Crosscutting Concepts |
|---|---|---|
| **Developing and Using Models**<br>Modeling in 9–12 builds on K–8 and progresses to using, synthesizing, and developing models to predict and explain relationships between systems and their components in the natural and designed world.<br>• Use multiple types of models to represent and support explanations of phenomena, and move flexibly between model types based on merits and limitations. (HS-LS4-b) | **LS4.B: Natural Selection**<br>• Natural selection occurs only if there is both (1) variation in the genetic information between organisms in a population and (2) variation in the expression of that genetic information—that is, trait variation—that leads to differences in performance among individuals. (HS-LS4-b),(HS-LS4-c)<br>**LS4.C: Adaptation**<br>• Natural selection is the result of four factors: (1) the potential for a species to increase in number, (2) the genetic variation of individuals in a species due to mutation and sexual reproduction, (3) competition for an environment's limited supply of the resources that individuals need in order to survive and reproduce, and (4) the ensuing proliferation of those organisms that are better able to survive and reproduce in that environment. (HS-LS4-b) | **Cause and Effect**<br>• Empirical evidence is required to differentiate between cause and correlation and make claims about specific causes and effects. (HS-LS4-b),(HS-LS4-d),(HS-LS4-e) |

Connections to other DCIs in this grade-level: will be added in future version.

Articulation of DCIs across grade-levels: will be added in future version.

Common Core State Standards Connections:

ELA /Literacy–

**RST.9-10.7** Translate quantitative or technical information expressed in words in a text into visual form (e.g., a table or chart) and translate information expressed visually or mathematically (e.g., in an equation) into words. (HS-LS4-b),(HS-LS4-c)

Mathematics –

**MP.4** Model with mathematics. (HS-LS4-b)

**F.LE** Construct and compare linear, quadratic, and exponential models and solve problems. ( HS-LS4-b)

When science teachers at all levels K–12 ask—"Where are the materials that help me teach to the standards?"—the educational system must have a concrete answer.

The instructional materials may be adapted from current programs, provided by states, or developed by organizations. They may come as hardback books,e-books, or other electronic forms; but, they must be available. At a minimum, model units are needed. Arguing for a coherent curriculum based on the standards is not new. Indeed, there is a long history of curriculum serving an essential role in science teaching. If there is no curriculum for teachers, I predict the standards will be implemented with far less integrity than intended by the *Framework* and those who developed the *NGSS.*

## Conclusion

The *NGSS* likely will influence K–12 science teaching for at least a decade, longer if recent history is any indication. This article uses the life sciences as the context for discussion of important content, some challenges, and several opportunities faced by K–12 teachers of science.

**HS-LS4 Biological Evolution: Unity and Diversity**

Students who demonstrate understanding can:

**HS-LS4-d. Construct an explanation based on evidence for how natural selection, genetic drift, gene flow through migration, and co-evolution lead to populations dominated by organisms that are anatomically, behaviorally, and physiologically adapted to survive and reproduce in a specific environment.** [Clarification Statement: Emphasis is on quantitative evidence as the basis for clarifying the difference among various processes of adaptation within populations. Data on specific environmental differences and selection for/against traits should be used. Environmental factors may include ranges of seasonal temperature, climate change, acidity, and light.] [Assessment Boundary: The assessment should measure students' abilities to differentiate types of evidence used in explanations.]

The performance expectations above were developed using the following elements from the NRC document *A Framework for K-12 Science Education*:

| Science and Engineering Practices | Disciplinary Core Ideas | Crosscutting Concepts |
|---|---|---|
| **Constructing Explanations and Designing Solutions**<br>Constructing explanations and designing solutions in 9–12 builds on K–8 experiences and progresses to explanations and designs that are supported by multiple and independent student-generated sources of evidence consistent with scientific knowledge, principles, and theories.<br>• Apply scientific reasoning, theory, and models to link evidence to claims to assess the extent to which the reasoning and data support the explanation or conclusion. (HS-LS4-d)<br>• Construct and revise explanations based on evidence obtained from a variety of sources (e.g., scientific principles, models, theories, simulations) and peer review. (HS-LS4-d)<br>• Base causal explanations on valid and reliable empirical evidence from multiple sources and the assumption that natural laws operate today as they did in the past and will continue to do so in the future. (HS-LS4-d) | **LS4.B: Natural Selection**<br>• The traits that positively affect survival are more likely to be reproduced, and thus are more common in the population. (HS-LS4-d),(HS-LS4-c),(HS-LS4-e)<br><br>**LS4.C: Adaptation**<br>• Natural selection leads to adaptation, that is, to a population dominated by organisms that are anatomically, behaviorally, and physiologically well suited to survive and reproduce in a specific environment. That is, the differential survival and reproduction of organisms in a population that have an advantageous heritable trait leads to an increase in the proportion of individuals in future generations that have the trait and to a decrease in the proportion of individuals that do not. (HS-LS4-d),(HS-LS4-c) | **Cause and Effect**<br>• Empirical evidence is required to differentiate between cause and correlation and make claims about specific causes and effects. (HS-LS4-b),(HS-LS4-d),(HS-LS4-e)<br>------------------------------<br>**Connections to Nature of Science**<br><br>**Scientific Knowledge Assumes an Order and Consistency (Regularity) in Natural Systems**<br>• Scientific knowledge is based on the assumption that natural laws operate today as they did in the past and they will continue to do so in the future. (HS-LS4-d),(HS-LS4-a) |

Connections to other DCIs in this grade-level: will be added in future version.

Articulation of DCIs across grade-levels: will be added in future version.

Common Core State Standards Connections:

ELA /Literacy–

**WHST.9-10.2** Write informative/explanatory texts, including the narration of historical events, scientific procedures/ experiments, or technical processes. (HS-LS4-d), (HS-LS4-e), (HS-LS4-a)

**WHST.9-10.4** Produce clear and coherent writing in which the development, organization, and style are appropriate to task, purpose, and audience. (HS-LS4-d), (HS-LS4-e), (HS-LS4-a)

**WHST.9-10.9** Draw evidence from informational texts to support analysis, reflection, and research. (HS-LS4-D), (HS-LS4-e), (HS-LS4-a)

**SL.9-10.2** Integrate multiple sources of information presented in diverse media or formats (e.g., visually, quantitatively, orally) evaluating the credibility and accuracy of each source. (HS-LS4-d),(HS-LS4-e),(HS-LS4-a)

*The NGSS life sciences team, co-chaired by Rodger Bybee and Brett Moulding, included contributions from the following individuals: Zoe Evans, Kevin Fisher, Jennifer Gutierrez, Chris Embry-Mohr, Julie Olson, Sherry Schaaf. Preliminary work for the National Research Council was compiled by Kathy Comfort, Danine Ezell, Bruce Fuchs, and Brian Reiser.

*Rodger W. Bybee* is a past executive director of the Biological Sciences Curriculum Study. He served as Design Team Lead for the NRC Framework and currently serves as the *NGSS* Writing Team Co-Leader for Life Sciences.

## References

American Association for the Advancement of Science (AAAS). 1993. *Benchmarks for science literacy.* Washington, DC: AAAS.

Biological Science Curriculum Study (BSCS). 1993. *Developing biological literacy.* Colorado Springs, CO: BSCS.

Bybee, R. W. 2011. The Next Generation Science Standards: Implications for high school biology.

*Research in Biology Education: Where Do We Go From Here?* Proceedings for a conference published by Michigan State University, Institute for Research on Mathematics and Science Education.

Bybee, R. W. 2012. The next generation of science standards: Implications for biology education. *The American Biology Teacher* 74 (8): 542–549.

Bybee, R., and M. Bloom, eds. 2008. *Measuring our success.* Dubuque, IA: Kendall/Hunt Publishing.

College Board. 2009. *Science college board standards for college success.* Available: *http://professionals.college board.com/profdownload/cbscs-science-standards-2009.pdf* [June 2011].

Hurd, P. D. 1961. *Biological education in American secondary schools 1890–1960.* Washington, DC: American Institute of Biological Sciences.

Kress, J., and G. Barrett. eds. 2001. *A new century of biology.* Washington, DC: Smithsonian Institution Press.

National Research Council (NRC). 1996. *National science education standards.* Washington, DC: National Academies Press.

National Research Council (NRC). 1999. *How people learn: Bridging research and practice.* Washington, DC: National Academies Press.

National Research Council (NRC). 2007. *Taking science to school: Learning and teaching science in grades K–8.* Washington, DC: National Academies Press.

National Research Council (NRC). 2009. *A new biology for the 21st century.* Washington, DC: National Academies Press.

National Research Council (NRC). 2012. *A framework for K–12 science education: Practices, crosscutting concepts, and core ideas.* Washington, DC: National Academies Press.

Wiggins, G., and J. McTighe. 2005. *Understanding by design.* Alexandria, VA: Association for Supervision and Curriculum Development (ASCD).

Willard, T., H. Pratt, and C. Workosky. 2012. Exploring the new standards. *The Science Teacher* 79 (7): 33–37.

# A Focus on Physical Science

*By Joseph Krajcik*

What should all students know about the physical sciences? Why should all students have a basic understanding of these ideas? An amazing number of new scientific breakthroughs have occurred in the last 20 years that impact our daily lives: genetics, nanoscience, and digital technologies, among many others. In addition, we have a much greater understanding of how students learn than ever before. With these breakthroughs, both in science and in how students learn science (NRC 2007), the National Research Council developed *A Framework for K–12 Science Education* (*Framework*; NRC 2012) to guide the development of the *Next Generation of Science Standards* (*NGSS*), scheduled for release this spring, that will provide direction in science teaching and learning. The overall goal of the *Framework* and *NGSS* is to help all learners in our nation develop the science and engineering understanding that they need to live successful, informed, and productive lives and that will help them create a sustainable planet for future generations. The physical science core ideas are critical to this effort.

The *NGSS* make use of five key ideas from the *Framework:* (1) limited number of core ideas, (2) crosscutting concepts, (3) engaging in scientific and engineering practices, (4) the integration or coupling of core ideas and scientific practices to develop performance expectations, and (5) an ongoing developmental process. The scientific and engineering practices and crosscutting concepts were discussed in the earlier chapters by Bybee and Duschl and are summarized in Figures 1 and 2.

**Figure 1. Practices for K–12 science curriculum**

- Asking questions (for science) and defining problems (for engineering).
- Developing and using models
- Planning and carrying out investigations
- Analyzing and interpreting data
- Using mathematics and computational thinking
- Constructing explanations (for science) and designing solutions (for engineering)
- Engaging in argument from evidence
- Obtaining, evaluating, and communicating information

**Figure 2. Crosscutting concepts for K–12 science education**

***Patterns.*** Observed patterns in nature guide organization and classification and prompt questions about relationships and causes underlying the patterns.

***Cause and effect: Mechanism and explanation.*** Events have causes, sometimes simple, sometimes multifaceted. Deciphering causal relationships and the mechanisms by which they are mediated is a major activity of science.

***Scale, proportion, and quantity.*** In considering phenomena, it is critical to recognize what is relevant at different sizes, times, and energy scales and to recognize proportional relationships between different quantities as scales change.

***Systems and system models.*** Delimiting and defining the system under study and making a model of it are tools for developing understanding used throughout science and engineering.

***Energy and matter: Flows, cycles, and conservation.*** Tracking energy and matter flows, into, out of, and within systems, helps one understand a system's behavior.

***Structure and function.*** The way an object is shaped or structured determines many of its properties and functions.

***Stability and change.*** For both designed and natural systems, conditions of stability and what controls rates of change are critical elements to consider and understand.

**Figure 3. Disciplinary core ideas in the physical sciences**

PS1: Matter and its interactions—How can one explain the structure, properties, and interactions of matter?

- PS1.A: Structure and Properties of Matter
- PS1.B: Chemical Reactions
- PS1.C: Nuclear Processes

PS2: Motion and stability: Forces and interactions—How can one explain and predict interactions between objects and within systems of objects?

- PS2.A: Forces and Motion
- PS2.B: Types of Interactions
- PS2.C: Stability and Instability in Physical Systems

PS3: Energy—How is energy transferred and conserved?

- PS3.A: Definitions of Energy
- PS3.B: Conservation of Energy and Energy Transfer
- PS3.C: Relationship Between Energy and Forces
- PS3.D: Energy in Chemical Processes and Everyday Life

PS4: Waves and Their Applications in Technologies for Information Transfer—How are waves used to transfer energy and information?

- PS4.A: Wave Properties
- PS4.B: Electromagnetic Radiation
- PS4.C: Information Technologies and Instrumentation

In this article, I will focus on the disciplinary core ideas in physical science, the development of those ideas across time, the importance of blending core ideas with scientific and engineering practices to build understanding, and the development of performance expectations.

The *Framework* and the *NGSS* focus on a limited number of core ideas of science and engineering both within and across the science disciplines that are essential to explain and predict a host of phenomena that students will encounter in their daily lives but that will also allow them to continue to learn more throughout their lives. Core ideas are powerful in that they are central to the disciplines of science, provide explanations of phenomena, and are the building blocks for learning within a discipline (Stevens, Sutherland, and Krajcik 2009). By focusing on ideas in depth, students learn the connections between concepts and principles so that they can apply their understanding to as yet unencountered situations, forming what is known as integrated understanding (Fortus and Krajcik 2011). Supporting students in learning integrated understanding is critical because it allows learners to solve real-world problems and to further develop understanding.

## The physical science core ideas

The core ideas in physical science will allow learners to answer important questions such as "How can we make new materials?" "Why do some things appear to keep going, but others stop?" and "How can information be shipped around wirelessly." Moreover, many phenomena, regardless of the discipline, require some level of understanding of physical and chemical ideas. An understanding of chemical reactions and the properties of elements and compounds serves as foundational knowledge for the life sciences and the Earth and space sciences. Explaining photosynthesis and respiration depends upon an understanding of chemical reactions. Understanding energy transfer is critical for explaining many phenomena in the life sciences and in the Earth and space sciences. Explaining ideas such as photosynthesis and plate tectonics depend on understanding

**Figure 4. A progression of ideas for the structure and properties of matter**

**By the end of second grade—a descriptive model:** Matter exists as different substances (e.g., wood, metal, water), and many of them can be either solid or liquid, depending on temperature. Substances can be described and classified by their observable properties (e.g., visual, aural, textural), by their uses, and by whether they occur naturally or are manufactured. Different properties are suited to different purposes. A great variety of objects can be built up from a small set of pieces. Objects or samples of a substance can be weighed, and their size can be described and measured.

**By the end of fifth grade—a particle model:** Matter of any type can be subdivided into particles that are too small to see, but even then the matter still exists and can be detected by other means (e.g., by weighing or by its effects on other objects). For example, a model showing that gases are made from matter particles that are too small to see and are moving freely around in space can explain many observations including the impacts of gas particles on surfaces (e.g., of a balloon) and on larger particles or objects (e.g., wind, dust suspended in air) and the appearance of visible scale water droplets in condensation, fog, and, by extension, also in clouds or the contrails of a jet. The amount (weight) of matter is conserved when it changes form, even in transitions in which it seems to vanish (e.g., sugar in solution, evaporation in a closed container). Measurements of a variety of properties can be used to identify particular substances.

**By the end of eighth grade—an atomic molecular model:** All substances are made from some 100 different types of atoms, which combine with one another in various ways. Atoms form molecules that range in size from two to thousands of atoms. Pure substances are made from a single type of atom or molecule; each pure substance has characteristic physical and chemical properties (for any bulk quantity under given conditions) that can be used to identify it. Gases and liquids are made of molecules or inert atoms that are moving about relative to each other. In a liquid, the molecules are constantly in contact with others; in a gas, they are widely spaced except when they happen to collide. In a solid, atoms are closely spaced and may vibrate in position but do not change relative locations. Solids may be formed from molecules, or they may be extended structures with repeating subunits. The changes of state that occur with variations in temperature or pressure can be described and predicted using these models of matter.

**By the end of twelfth grade—an atomic structure model:** Each atom has a charged substructure consisting of a nucleus, which is made of protons and neutrons, surrounded by electrons. The periodic table orders elements horizontally by the number of protons in the atom's nucleus and places those with similar chemical properties in columns. The repeating patterns of this table reflect patterns of outer electron states. The structure and interactions of matter at the bulk scale are determined by electrical forces within and between atoms. Stable forms of matter are those in which the electric and magnetic field energy is minimized. A stable molecule has less energy, by an amount known as the binding energy, than the same set of atoms separated; one must provide at least this energy to take the molecule apart.

(Adapted from *A Framework for K–12 Science Education: Practices, Crosscutting Concepts, and Core Ideas* [NRC 2012])

of energy transfer. Explaining how the planets revolve around the Sun depends on understanding gravitational force. Explaining why some materials are attracted to each other while others are not depends upon an understanding of electrical forces. Being able to explain why earthquakes can cause so much damage depends on an understanding of energy transfer. As such, a major goal of the *Framework* is for students to see that the underlying cause-and-effect relationships that occur in all systems and processes, whether biological or physical, can be understood through physical and chemical processes. Because the physical science ideas explain many natural and human-made phenomena that occur each day, developing integrated understanding of them is important for all learners and not only those going on to study science in college or interested in a career in science.

The *Framework* identifies four core ideas in physical science—a blending of chemistry and physics. Figure 3 (p. 114) presents a list of these core ideas.

### Core Idea 1: PS1: Matter and Its Interactions

The first core idea, PS1: Matter and Its Interactions, helps students to formulate an answer to the question"How can one explain the structure, properties, and interactions of matter?" Understanding matter, its properties, and how it undergoes changes is critical to explaining phenomena in physical science and in the life, and Earth and space sciences. This core idea explains phenomena such as a puddle of water evaporating, burning of wood, tarnishing of metal statues, the cycling of carbon in the environment, and why so many diverse and new products can be formed from such a small set of elements. Although the periodic table identifies 118 elements, only a quarter of these are responsible for all the products on Earth; and fewer than 10, including carbon, hydrogen, oxygen, and nitrogen, make up most materials. These materials exist because in chemical reactions, while the various types and number of atoms are conserved, the arrangement of the atoms is changed, explaining the many observable phenomena in living and nonliving systems.

### Core Idea 2: PS2: Motion and Stability

"How can one explain and predict interactions between objects and within systems of objects?" The second core idea, PS2: Motion and Stability: Forces and Interactions, focuses on helping students understand ideas related to why some objects will keep moving, why objects fall to the ground, and why some materials are attracted to each other while others are not. Supporting students in developing an understanding of the forces between objects is important for describing and explaining how the motion of objects change, as well as for predicting stability or instability in systems at any scale. The *Framework* describes the forces between objects arising from a few types of interactions: gravity, electromagnetism, and the strong and weak nuclear interactions. The *Framework* places an emphasis on these forces being explained by force fields that contain energy that can transfer energy through space. The *Framework,* while not ignoring gravitational forces, places equal weight on helping students understand electrical interactions as the force that holds various materials together. The attraction and repulsion of electric charges at the atomic scale provide an explanation for the structure, properties, and transformations of matter. Although the ideas of force fields and electrical interactions aren't new, their emphasis as critical to explain everyday phenomena is.

### Core Idea 3: PS3: Energy

The third core idea, PS3: Energy, answers the question "How is energy transferred and conserved?" Energy, while difficult to define, explains the interactions of objects using the ideas of transfer of energy from one object or system of objects to another and that energy is always conserved. How is it that power plants can provide energy used to run household appliances? Understanding energy transfer is critical to this idea. Equally important is for students to understand that the total energy within a defined system changes only by transferring energy into or out of the system with the total amount of energy remaining constant—the conservation of energy. Although energy is always conserved, it can be converted to less useful forms, such as thermal energy in the surroundings. Energy transfer and conservation are critical ideas to explain diverse phenomena such as photosynthesis, respiration, plate tectonics, combustion, and various energy storage devices, such as batteries. While all disciplines have energy as an important construct, often energy is not well understood by students.

### Core Idea 4: PS4: Waves and Their Applications in Technologies for Information Transfer

The fourth core idea, PS4: Waves and Their Applications in Technologies for Information Transfer, is critical to understanding how many new technologies work and how information is shipped around and stored. This core idea introduces students to critical ideas that explain how the sophisticated technologies available today and how various forms of light and sound are mechanisms for the transfer of energy and transfer of information among objects not in contact with each other. As such, this core idea helps to answer the question: "How are waves used to transfer energy and send and store information?" This core idea also stresses the interplay of physical science and technology. Modern communication, information, and imaging technologies are pervasive in our lives today and serve as critical tools that scientists use to explore the many scales that humans could not explore without these tools. Understanding how these pervasive tools work requires that we understand light and sound and their interactions with matter.

## Learning develops over time

The *Framework* goes beyond just presenting the final endpoint for each core idea. Rather, the document is structured with grade band endpoints, consistent with what is known about how learning occurs as an ongoing developmental process. A developmental perspective purposefully builds and links to students' current understanding to form richer and more connected ideas over time (NRC 2007). The core ideas discussed above should be developed from elementary through high school as each year student ideas become more sophisticated, allowing them to more completely explain phenomena as well as explain more phenomena. Too often in science education, we have not systematically considered the prior knowledge of children to build deep and more connected understanding from kindergarten through high school; to do so is critical to build understanding that can be used to solve problems.

A developmental approach guides students' knowledge toward a more sophisticated and coherent understanding of the scientific idea (NRC 2007; Corcoran, Mosher, and Rogat 2009). At the elementary level students explore ideas at an experiential level. For instance, they explore which type of materials that they experience can be melted or turned into a solid. Although learners continue to experience phenomena as they continue in their school, ideas that explain these ideas are introduced. Students form a solid grasp of a particle model in fifth or sixth grade to explain phase changes and then refine this model in seventh and eighth grade so they know that the particles are made of atoms or molecules to explain and predict even more complex phenomena such as chemical reactions would be an example of a developmental approach.

Grade band endpoints in *A Framework for K–12 Science Education* show an indication of this progression of ideas across time. As such, the *Framework* presents a coherent picture of how ideas should develop across time. Figure 4 (p. 115) shows a progression for the core idea Structure and Properties of Matter. At each grade band, students develop a conceptual model that they can use to explain phenomena. At the second grade level, students develop a descriptive model that they can use to describe how matter can exist in different phases. As they continue with their schooling, their conceptual model becomes more sophisticated. By the end of secondary school, students have developed an atomic structure model that allows them to use a causal model for explaining the structure of matter.

This growth in understanding is not developmentally inevitable, but depends upon instruction and key learning experiences to support students in developing more sophisticated understanding across time. Reaching these endpoints depends upon the instruction the student receives and how understanding is assessed. To be a complete learning progression, the progression would also need to show how you can move students from one level to the next and how to assess that understanding. These instructional components are not part of the *Framework* but will depend on development of new curriculum materials based on research. (For more on learning progression research, see Smith et al. 2006 and Rogat et al. 2011).

## Content (scientific ideas) is not enough!

The *Framework,* however, stresses more than just ideas in the disciplines. The *Framework* also presents the scientific and engineering practices and crosscutting concepts that students need to use in conjunction with core ideas to build understanding. Scientific practices consist of the multiple ways in which science explores and understands the world (see "Scientific and Engineering Practices in K–12 Classrooms" by Bybee [p.39]). Crosscutting concepts are major ideas that transverse the various scientific disciplines (see "The Second Dimension—Crosscutting Concepts" by Duschl [p. 57]). The *Framework* emphasizes that learning about science and engineering involves the coupling of core ideas with scientific and engineering practices and crosscutting concepts to engage students in scientific inquiry and engineering design. Convincing evidence exists that understanding science will only result when core ideas are blended with scientific and engineering practices and crosscutting concepts (NRC 2007). Just as science is both a body of knowledge and the process whereby that body of knowledge is developed, the learning of science is similar: You cannot learn a core idea without using it with scientific or engineering practices. Therefore, using practices as a means to develop understanding of science ideas should be a regular part of students' classroom experience and is emphasized throughout the *Framework.*

## Expressing standards as performance expectations

The *Framework* stresses that standards should emphasize all three dimensions by integrating scientific and engineering practices with crosscutting concepts and disciplinary core ideas to

**Figure 5. Sample standards for different grade levels**

Sample standards in physical science—kindergarten

**K-PS1 Matter and Its Interactions**

Students who demonstrate understanding can:

**K-PS1-b. Design and conduct investigations to test the idea that some materials can be a solid or liquid depe temperature.** [Assessment Boundary: Only a qualitative description of temperature should be used such as hot, cool, and warm.]

The performance expectations above were developed using the following elements from the NRC document *A Framework for K-12 Science Educatio*

| Science and Engineering Practices | Disciplinary Core Ideas | Crosscutting Conc |
|---|---|---|
| **Planning and Carrying Out Investigations**<br>Planning and carrying out investigations to answer questions or test solutions to problems in K–2 builds on prior experiences and progresses to simple investigations, based on fair tests, which provide data to support explanations or design solutions.<br>• With guidance, design and conduct investigations in collaboration with peers. (K-PS1-a), (K-PS1-b)<br>• Make direct or indirect observations and/or measurements to collect data which can be used to make comparisons. (K-PS1-a), (K-PS1-b)<br>---<br>**Connections to Nature of Science**<br>**Science Knowledge is Based on Empirical Evidence**<br>• Scientists look for patterns and order when making observations about the world. (K-PS1-a), (K-PS1-b), (K-PS1-c) | **PS1.A: Structure and Properties of Matter**<br>• Different kinds of matter exist (e.g., wood, metal, water) and many of them can be either solid or liquid, depending on temperature. (K-PS1-a), (K-PS1-b) | **Cause and Effect**<br>• Events have causes that generat patterns. (K-PS1-b)<br>• Simple tests can be designed to evidence to support or refute stu about causes. (K-PS1-b) |

Connections to other DCIs in this grade-level: will be added in future version.

Articulation of DCIs across grade-levels: will be added in future version.

Common Core State Standards Connections:

Mathematics –

**MP.3** Construct viable arguments and critique the reasoning of others. (K-PS1-b)

**K.MD.1** Describe measurable attributes of objects, such as length or weight. Describe several measurable attributes of a single object. (K-PS1-a),(K-PS1-b)

**K.MD.2** Directly compare two objects with a measurable attribute in common, to see which object has "more of"/"less of" the attribute, and describe the differer PS1-a),(K-PS1-b)

Sample standards in physical science—grade 2

**2. PS1 Matter and Its Interactions**

Students who demonstrate understanding can:

**2-PS1-d. Identify arguments that are supported by evidence that some changes caused by heating or cooling can be reversed and some cannot.** [Clarification Statement: Examples of reversible changes are melting chocolate or freezing liquids. An irreversible change is cooking food.]

The performance expectations above were developed using the following elements from the NRC document *A Framework for K-12 Science Education*.

| Science and Engineering Practices | Disciplinary Core Ideas | Crosscutting Concepts |
|---|---|---|
| **Engaging in Argument from Evidence**<br>Engaging in argument from evidence in K–2 builds on prior experiences and progresses to comparing ideas and representations about the natural and designed world.<br>• Identify arguments that are supported by evidence. (2-PS1-d)<br>---<br>**Connections to Nature of Science**<br>**Science Models, Laws, Mechanisms, and Theories Explain Natural Phenomena**<br>• Science searches for cause and effect relationships to explain natural events. (2-PS1-d) | **PS1.B: Chemical Reactions**<br>• Heating or cooling a substance may cause changes that can be observed. Sometimes these changes are reversible (e.g., melting and freezing), and sometimes they are not (e.g., baking a cake, burning fuel). (2-PS1-d) | **Scale, Proportion, and Quantity**<br>• Relative scales allow objects to be compared and described (e.g., bigger and smaller; hotter and colder; faster and slower). (2-PS1-d) |

Connections to other DCIs in this grade-level: will be added in future version.

Articulation of DCIs across grade-levels: will be added in future version.

Common Core State Standards Connections:

ELA/Literacy –

**RI.2.8** Explain how an author uses reasons and evidence to support particular points in a text, identifying which reasons and evidence support which point(s). (2-PS1-d)

**RI.2.10** By the end of year, read and comprehend informational texts, including history/social studies, science, and technical texts, in the grades 2–3 text complexity band proficiently, with scaffolding as needed at the high end of the range. (2-PS1-c), (2-PS1-d), (2-PS1-a)

**W.2.8** Describe how reasons support specific points the author makes in a text. (2-PS1-d)

Mathematics –

**MP.3** Construct viable arguments and critique the reasoning of others. (2-PS1-d)

**Figure 5. (*continued*)**

Sample standards in physical science—grade 5

**5-PS1 Matter and Its Interactions**

Students who demonstrate understanding can:

**5-PS1-d. Design and conduct investigations on the mixing of two or more different substances to determine whether a new substance with new properties is formed.** [Clarification Statement: Examples of interactions forming new substances can include mixing baking soda and vinegar. Examples of interactions not forming new substances can include mixing baking soda and water.]

The performance expectations above were developed using the following elements from the NRC document *A Framework for K-12 Science Education*:

| Science and Engineering Practices | Disciplinary Core Ideas | Crosscutting Concepts |
|---|---|---|
| **Planning and Carrying Out Investigations**<br>Planning and carrying out investigations to answer questions or test solutions to problems in 3–5 builds on K–2 experiences and progresses to include investigations that control variables and provide evidence to support explanations or design solutions.<br>• Design and conduct investigations collaboratively, using fair tests in which variables are controlled and the number of trials considered. (5-PS1-d)<br>• Make observations and/or measurements, collect appropriate data, and identify patterns that provide evidence for an explanation of a phenomenon or test a design solution. (5-PS1-c),(5-PS1-d) | **PS1.B: Chemical Reactions**<br>• When two or more different substances are mixed, a new substance with different properties may be formed; such occurrences depend on the substances and the temperature. (5-PS1-d),(5-PS1-e) | **Cause and Effect**<br>• Cause and effect relationships are routinely identified, tested, and used to explain change. (5-PS1-d),(5-PS1-e) |

Connections to other DCIs in this grade-level: will be added in future version.

Articulation of DCIs across grade-levels: will be added in future version.

Common Core State Standards Connections:

ELA/Literacy –

**W.5.7** Conduct short research projects that use several sources to build knowledge through investigation of different aspects of a topic. (5-PS1-c),(5-PS1-d)

Mathematics –

**MP.2** Reason abstractly and quantitatively. (5-PS1-d),(5-PS1-b)

**5.OA.2** Write and interpret numerical expressions. (5-PS1-d),(5-PS1-e)

**4.MD.2** Use the four operations to solve word problems involving distances, intervals of time, liquid volumes, masses of objects, and money, including problems involving simple fractions or decimals, and problems that require expressing measurements given in a larger unit in terms of a smaller unit. Represent measurement quantities using diagrams such as number line diagrams that feature a measurement scale. (5-PS1-d),(5-PS1-e)

Sample standards in physical science—middle school

**MS-PS1 Matter and Its Interactions**

Students who demonstrate understanding can:

**MS-PS1-d. Develop molecular models of reactants and products to support the explanation that atoms, and therefore mass, are conserved in a chemical reaction.** [Clarification Statement: Models can include physical models and drawings that represent atoms rather than symbols. The focus is on law of conservation of matter.] [Assessment Boundary: The use of atomic masses is not required. Balancing symbolic equations (e.g. N2 + H2 -> NH3) is not required.]

The performance expectations above were developed using the following elements from the NRC document *A Framework for K-12 Science Education*:

| Science and Engineering Practices | Disciplinary Core Ideas | Crosscutting Concepts |
|---|---|---|
| **Developing and Using Models**<br>Modeling in 6–8 builds on K–5 and progresses to developing, using and revising models to support explanations, describe, test, and predict more abstract phenomena and design systems.<br>• Use and/or develop models to predict, describe, support explanation, and/or collect data to test ideas about phenomena in natural or designed systems, including those representing inputs and outputs, and those at unobservable scales. (MS-PS1-a), (MS-PS1-c), (MS-PS1-d)<br>---------------------------------------------<br>**Connections to Nature of Science**<br>**Science Models, Laws, Mechanisms, and Theories Explain Natural Phenomena**<br>• Laws are regularities or mathematical descriptions of natural phenomena. (MS-PS1-d) | **PS1.B: Chemical Reactions**<br>• Substances react chemically in characteristic ways. In a chemical process, the atoms that make up the original substances are regrouped into different molecules, and these new substances have different properties from those of the reactants. (MS-PS1-d), ( MS-PS1-e), (MS-PS1-f)<br>• The total number of each type of atom is conserved, and thus the mass does not change. (MS-PS1-d) | **Energy and Matter**<br>• Matter is conserved because atoms are conserved in physical and chemical processes. (MS-PS1-d) |

Connections to other DCIs in this grade-level: will be added in future version.

Articulation of DCIs across grade-levels: will be added in future version.

Common Core State Standards Connections:

ELA/Literacy –

**RST.6-8.7** Integrate quantitative or technical information expressed in words in a text with a version of that information expressed visually (e.g., in a flowchart, diagram, model, graph, or table). (MS-PS1-c), (MS-PS1-d),(MS-PS1-g)

Mathematics –

**MP.9** Look for and express regularity in repeated reasoning. (MS-PS1-d)

**Figure 5. (*continued*)**

Sample standards in physical science—high school

**HS-PS1 Matter and Its Interactions**

Students who demonstrate understanding can:

**HS-PS1-i. Construct an explanation to support predictions about the outcome of simple chemical reactions, using the structure of atoms, trends in the periodic table, and knowledge of the patterns of chemical properties.** [Clarification Statement: Examples of chemical reactions would include the reaction of sodium and chlorine, or carbon and oxygen, or carbon and hydrogen.] [Assessment Boundary: Chemical reactions not readily predictable from the element's position on the periodic table (i.e., the main group elements) and combustion reactions are not intended. Reactions typically classified by surface level characteristics (e.g., double or single displacement reactions) are not intended.]

The performance expectations above were developed using the following elements from the NRC document *A Framework for K-12 Science Education*.

| Science and Engineering Practices | Disciplinary Core Ideas | Crosscutting Concepts |
|---|---|---|
| **Constructing Explanations and Designing Solutions** Constructing explanations and designing solutions in 9–12 builds on K–8 experiences and progresses to explanations and designs that are supported by multiple and independent student-generated sources of evidence consistent with scientific knowledge, principles, and theories.<br>• Construct and revise explanations based on evidence obtained from a variety of sources (e.g., scientific principles, models, theories, simulation) and peer review. (HS-PS1-e),(HS-PS1-i)<br>--------------------------------------------<br>**Connections to Nature of Science**<br>**Science Models, Laws, Mechanisms, and Theories Explain Natural Phenomena**<br>• Laws are regularities or mathematical descriptions of natural phenomena. (MS-PS1-d) | **PS1.B: Chemical Reactions**<br>• The fact that atoms are conserved, together with knowledge of the chemical properties of the elements involved, can be used to describe and predict chemical reactions. (HS-PS1-h),(HS-PS1-i) | **Patterns**<br>• Different patterns may be observed at each of the scales at which a system is studied and can provide evidence for causality in explanations of phenomena. (HS-PS1-i) |

Connections to other DCIs in this grade-level: will be added in future version.

Articulation of DCIs across grade-levels: will be added in future version.

Common Core State Standards Connections:

ELA/Literacy –

| | |
|---|---|
| **RST.9-10.1** | Cite specific textual evidence to support analysis of science and technical texts, attending to the precise details of explanations or descriptions. (HS-PS1-e), (HS- PS1-i) |
| **RST.9-10.9** | Compare and contrast findings presented in a text to those from other sources (including their own experiments), noting when the findings support or contradict previous explanations or accounts. (HS-PS1-e), (HS-PS1-i) |
| **RST.11-12.9** | Synthesize information from a range of sources (e.g., texts, experiments, simulations) into a coherent understanding of a process, phenomenon, or concept, resolving conflicting information when possible. (HS-PS1-e), (HS-PS1-i) |
| **WHST.11-12. 2** | Write informative/explanatory texts, including the narration of historical events, scientific procedures/ experiments, or technical processes. (HS-PS1-e), (HS-PS1-i) |
| **WHST.11-12.4** | Produce clear and coherent writing in which the development, organization, and style are appropriate to task, purpose, and audience. (HS-PS1-a), (HS-PS1-e),(HS- PS1-i) |
| **WHST.9-10.9** | Draw evidence from informational texts to support analysis, reflection, and research. (HS-PS1-e),(HS-PS1-i) |
| **SL.9-10.2** | Integrate multiple sources of information presented in diverse media or formats (e.g., visually, quantitatively, orally) evaluating the credibility and accuracy of each source. (HS-PS1-a), (HS-PS1-c), (HS-PS1-e), (HS-PS1-i) |

Mathematics –

| | |
|---|---|
| **S.IC.B** | Make inferences and justify conclusions from sample surveys, experiments, and observational studies. (HS-PS1-b), (HS-PS1-e), (HS-PS1-h), (HS-PS1-i) |

develop performance expectations. Performance expectations require that students demonstrate knowledge-in-use (NRC 2012). As such, *NGSS* in physical science will be written in terms of performance expectations. Figure 5 (pp.119–121) shows five possible performance expectations related to PS1: Matter and Its Interactions. These five performance expectations blend the core ideas with scientific practices and crosscutting concepts—and suggest that learning across the grades become more sophisticated. Examining the performance expectations, the idea of chemical reactions becomes more sophisticated from elementary to high school, and it allows learners to explain and predict more phenomena.

## Concluding thought

Because fewer ideas are presented and developed across K–12 science curriculum and blended with the use of scientific practices and crosscutting elements, the *NGSS* will present a more coherent view of science education. By developing understanding of the physical science core ideas, students will develop responses to three critical questions: "What is everything made of?" "Why do things happen?" and "How are waves used to transfer energy and information?" Being able to answer these questions will provide students with the conceptual tools to explain phenomena, solve problems, and learn more as needed. Students will begin to build understanding to these questions in the early elementary grades and will continue their development through high school.

The *Framework* and *NGSS* also emphasize the blending of "content" and "inquiry" to build understanding. The performance expectations in the *NGSS* are the endpoints that learners will need to meet. Classroom instruction and curriculum materials will need to not only help students reach these important ideas but also involve learners in using scientific practices blended with the core ideas and crosscutting concepts to develop and apply the scientific ideas. The core ideas and the performance expectations in physical science are especially important as they build foundational ideas for explaining phenomena in other disciplines.

*Joe Krajcik* is a professor of science education at Michigan State University and director of the Institute for Collaborative Research for Education, Assessment, and Teaching Environment for Science, Technology, and Engineering and Mathematics (CREATE for STEM). He served as design team lead for the NRC *Framework* and serves as design team lead for *NGSS*.

### Acknowledgments

Thanks to Ann Novak, middle school science teacher from Greenhills School in Ann Arbor, Michigan, for her helpful comments on this manuscript.

## References

Corcoran, T., F. A. Mosher, and A. Rogat. 2009. *Learning progressions in science: An evidence-based approach to reform.* Center on Continuous Instructional Improvement. New York: Teachers College, Columbia University.

Fortus, D., and J. S. Krajcik. 2011. Curriculum coherence and learning progressions. In *Second international handbook of science education,* ed. B. J. Fraser, K. G. Tobin, and C. J. McRobbie, pp. 783–798. Dordrecht, Netherlands: Springer.

National Research Council (NRC). 2007. *Taking science to school: Learning and teaching science in grades K–8.* Washington, DC: National Academies Press.

National Research Council (NRC). 2012. *A framework for K–12 science education: Practices, crosscutting concepts, and core ideas.* Washington, DC: National Academies Press.

Rogat, A., C. Anderson, J. Foster, F. Goldberg, J. Hicks, D. Kanter, J. Krajcik, R. Lehrer, B. Reiser, and M. Wiser. 2011. Developing learning progressions in support of the new science standards: A RAPID workshop series. CPRE: University of Pennsylvania. *www.cpre.org/developing-learning-progressions-support-new-science-standards-rapid-workshop-series.*

Smith, C. L., M. Wiser, C. W. Anderson, and J. Krajcik. 2006. Implications of research on children's learning for standards and assessment: A proposed learning progression for matter and the atomic molecular theory. *Measurement: Interdisciplinary Research and Perspectives* 14 (1 and 2): 1–98.

Stevens, S. Y., L. M. Sutherland, and J. Krajcik. 2009. *The big ideas of nanoscale science and engineering: A guidebook for secondary teachers.* Arlington, VA: NSTA Press.

# The *Next Generation Science Standards* and the Earth and Space Sciences

***By Michael E. Wysession***

The *Next Generation Science Standards (NGSS)* represent a revolutionary step toward establishing modern national K–12 science education standards. Based on the recommendations of the National Research Council's *A Framework for K–12 Science Education: Practices, Crosscutting Concepts, and Core Ideas* (*Framework*; NRC 2012), these performance expectations present a progressive approach to developing students' understanding of science. The writing of the *NGSS* has been supervised by Achieve Inc., the bipartisan not-for-profit organization that also supervised the writing of the math and English language arts (ELA) *Common Core State Standards* (NGAC and CCSSO 2010). The *NGSS* involve significant changes from traditional standards at all levels for all of the sciences, integrating three dimensions of science content, science practices, and the crosscutting, big-picture themes of science. Nowhere are these changes more apparent than for the Earth and space sciences (ESS), which now require a year of upper-level high school coursework.

## Emphasis on Earth and space sciences

Earth and space sciences have a strong presence within the *NGSS,* especially in high school. Earth and space sciences traditionally occupy about a third of the middle school science curriculum, on par with life and physical science. This parity often breaks down in high school with science curricula dominated by physics, chemistry, and biology. The *NGSS* represent a major departure from this, with the number of performance expectations in Earth and space sciences roughly equal to that of both life and physical sciences. At a high school level, the amount of content in Earth and space sciences is roughly equal to the amount of chemistry and physics combined.

For Earth and space sciences, within both the *Framework* and *NGSS,* there is a significant change toward a systems-approach that draws upon the wealth of Earth systems science research in both content and pedagogy (e.g., Ireton, Mogk, and Manduca 1996). Within the *Framework,* Earth and space science content is parsed into three big ideas, each of which is subdivided into components (Figure 1, p. 124). ESS1 examines the space and solar *systems;* ESS2 examines the interconnections among Earth's many different *systems* of the geosphere, hydrosphere, atmosphere, cryosphere, and biosphere; and ESS3 focuses on the anthroposphere *system,* the important role that human civilization plays in affecting Earth's other systems.

The geosphere is Earth's rock and metal; the hydrosphere is Earth's water; the atmosphere is Earth's air; the cryosphere is Earth's ice; and the biosphere is Earth's life. These overlap significantly. For example, the atmosphere contains parts of all the other spheres in the form of dust, ice, water, and living organisms.

## Motivation for Earth and space sciences

The third big idea—human interactions—is the primary reason for the increased attention given to Earth and space sciences in the *NGSS*. This big idea contains many of the most newsworthy topics in science: natural hazards, energy sources, water and mineral availability, climate change, environmental impacts, and human sustainability.

Perhaps the greatest change in Earth and space science content from the influential *National Science Education Standards* (NRC 1996) comes from the increased awareness of the enormous magnitude of the effects of human activities on our planet. Humans now use almost 40% of Earth's land surface to produce food. The land area devoted to roads and parking lots in the United States is larger than the state of Georgia. Americans use roughly 4 billion metric tons of non–energy-related rocks and minerals each year (~10× the total material carried by the entire Mississippi River system) to construct all the material objects of our lives, roughly 25,000 pounds per person. The release of acidic aerosols in the atmosphere

**Figure 1. The Big Ideas of the *Framework* and standard topics of the *NGSS***

The Big Ideas and subtopics of the *Framework* (NRC 2012) are shown on the left below. The standard topics for high school and middle school from the January 2013 public draft of the *NGSS* (Achieve Inc.) are on the right. The number of standard topics differs for middle and high school because the middle school standards for *Interior Systems* and *Surface Systems* are merged into a single high school *Earth's Systems*. Elementary is not shown because of the high level of integration of science content among all of the sciences in grades K–5.

| ***Framework for K–12 Science Education*** | ***Next Generation Science Standards: Middle School – Earth and Space Science Topics*** (Draft, January 2013) |
|---|---|
| ESS1: Earth's Place in the Universe | MS.Space Systems |
| ESS1.A: The Universe and Its Stars | MS.History of Earth |
| ESS1.B: Earth and the Solar System | MS.Earth's Interior Systems |
| ESS1.C: The History of Planet Earth | MS.Earth's Surface Systems |
| ESS2: Earth's Systems | MS.Weather and Climate |
| ESS2.A: Earth Materials and Systems | MS.Human Impacts |
| ESS2.B: Plate Tectonics and Large-Scale System interactions | ***Next Generation Science Standards: High School – Earth and Space Science Topics*** (Draft, January 2013) |
| ESS2.C: The Roles of Water in Earth's Surface Processes | HS.Space Systems |
| ESS2.D: Weather and Climate | HS.History of Earth |
| ESS2.E: Biogeology | HS.Earth's Systems |
| ESS3: Earth and Human Activity | HS.Climate Change |
| ESS3.A: Natural Resources | HS.Human Sustainability |
| ESS3.B: Natural Hazards | |
| ESS3.C: Human Impacts on Earth Systems | |
| ESS3.D: Global Climate Change | |

have increased global land erosion rates and have increased ocean acidity by 30% in just a few centuries.

Human impacts are no longer an asterisk in Earth science: Our activities are changing the composition of the atmosphere, hydrosphere, biosphere, and cryosphere and altering land surfaces faster than any other natural process. In fact, the sphere of human impact on Earth systems—the *anthroposphere*—is now the greatest agent of geologic change on our planet's surface, and this needs to be reflected in our curricula. The *NGSS* pass no moral judgment on these human activities. The impacts are simply the reality of the immense power of our species and need to be recognized as such in our educational standards.

## Connections across and between standards

The *NGSS* strive for a greater integration among the sciences and engineering, as well as with the *Common Core State Standards*. This integration reinforces concepts across fields to enrich

**Figure 2. Essentials of *A Framework for K–12 Science Education***

The *Framework* (NRC 2012) presents fundamental concepts and practices for the new standards and implied changes in K–12 science programs. The *Framework* describes three essential dimensions: science and engineering practices, crosscutting concepts, and core ideas in science disciplines. In this article, the core disciplinary ideas are from the Earth and space sciences.

The scientific and engineering practices have been discussed in earlier chapters and are summarized below.

**Practices for K–12 science curriculum**

1. Asking questions (for science) and defining problems (for engineering)
2. Developing and using models
3. Planning and carrying out investigations
4. Analyzing and interpreting data
5. Using mathematics and computational thinking
6. Constructing explanations (for science) and designing solutions (for engineering)
7. Engaging in argument from evidence
8. Obtaining, evaluating, and communicating information

The second dimension described in the NRC *Framework* is crosscutting concepts. These too have been discussed in earlier articles and are summarized here.

**Crosscutting concepts for K–12 science education**

1. *Patterns.* Observed patterns in nature guide organization and classification and prompt questions about relationships and causes underlying the patterns.
2. *Cause and effect: Mechanism and explanation.* Events have causes, sometimes simple, sometimes multifaceted. Deciphering causal relationships and the mechanisms by which they are mediated is a major activity of science.
3. *Scale, proportion, and quantity.* In considering phenomena, it is critical to recognize what is relevant at different sizes, times, and energy scales and to recognize proportional relationships between different quantities as scales change.
4. *Systems and system models.* Delimiting and defining the system under study and making a model of it are tools for developing understanding used throughout science and engineering.
5. *Energy and matter: Flows, cycles, and conservation.* Tracking energy and matter flows, into, out of, and within systems, helps one understand a system's behavior.
6. *Structure and function.* The way an object is shaped or structured determines many of its properties and functions.
7. *Stability and change.* For both designed and natural systems, conditions of stability and what controls rates of change are critical elements to understand.

and deepen students' understanding. The divisions among scientific fields is entirely arbitrary; nature knows or cares nothing about biology, chemistry, geology, or physics; there is only nature. Earth and space sciences, in particular, are the most integrated of all the sciences; most faculty members in university Earth science departments would not identify themselves as geologists but rather as geochemists, geophysicists, or geobiologists. Better integration among the sciences is required for the optimal teaching of Earth and space sciences.

One way the *NGSS* better integrate the sciences is with crosscutting concepts, which are universal principles (e.g., *cause and effect, structure and function)* that apply to all sciences. Each performance expectation of each standard topic is tied to the most closely related crosscutting concept. Standard topics often have strong connections with one particular crosscutting concept. For example, the middle school *Space Systems* topic connects well with the crosscutting concept of *scale, proportion, and quantity,* and the high school *Earth's Systems* topic connects well with the crosscutting concept of *matter and energy: flows and cycles.* The *NGSS* performance expectations are also aligned with concepts of the nature of science (Achieve Inc. 2013, Appendix H), which are integrated with both the science and engineering practices and the crosscutting concepts and therefore can appear in either one of the associated foundation boxes.

There is a greater connection within the *NGSS* between the basic sciences and the fields of engineering and technology, recognizing that a continuum rather than a sharp line exists between science and engineering. Wherever possible, the standards have incorporated aspects of Engineering, Technology, and the Applications of Science. This has been easy for Earth and space sciences because of the emphasis on human-related areas of natural resources, hazards, and human impacts. The *NGSS* Earth and space science topics of Human Impacts in middle school and Human Sustainability in high school are primarily focused on how human technology is affecting Earth's other systems and how that same technology can also be used to monitor, understand, and minimize these impacts. However, engineering and technology connections play important roles throughout the Earth and space sciences. For example, NASA technology provides the basis for our understanding of the space and solar systems, and weather predictions such as the path and timing of Hurricane Sandy are only possible because of the complex monitoring systems that continuously gather weather data and the computer models that analyze them.

The *NGSS* are also correlated with grade-appropriate *Common Core State Standards* in both mathematics and ELA. Reading and writing in the sciences pose unique ELA challenges. The Earth and space sciences involve concepts and vocabularies that are often new and unfamiliar to students, and both the science and language education benefit from the coordination of their progressions. ELA standards are especially connected to the *NGSS* through the practice of "Obtaining, Evaluating, and Communicating Information."

At a research level, the Earth and space sciences are as mathematical, quantitative, and computational as any other field of study, but that has not been adequately represented in K–12 education, where geoscience is often still presented as a field of qualitative categorization. For example, identifying minerals or types of rock has little value by itself but is important in understanding complex quantitative processes in the cycling of matter among Earth systems. Weather and climate are often taught as exercises in memorizing categories of cloud or climate types; in the *NGSS*, however, weather, climate, and climate change are presented as dynamic sciences that are mathematically and computationally complex, and inasmuch as climate incorporates all of Earth's systems (including the solar system) and requires significant understanding of physical and life sciences, it can be seen as a capstone high school experience that integrates all of the sciences in a quantitative and societally relevant experience.

**Figure 3. The *Framework***

The *Framework for K–12 Science Education* (NRC 2012), funded by the Carnegie Foundation of New York, provides the blueprint for a comprehensive set of K–12 science standards for life, physical, and Earth and space sciences. The *Framework*, publicly available (*www.nap.edu*), draws upon the best practices of pedagogy and methods for K–12 science education (NRC 2007, 2008) and advocates construction of science standards by interweaving three distinct aspects, or dimensions: (1) the practices of science, (2) the disciplinary core ideas of science, and (3) the broad crosscutting concepts of science. The ideas of this blueprint were integrated into performance expectations for the *NGSS*.

## Elementary school

At the elementary level, the *NGSS* present performance expectations that are integrated across sciences and organized into broad topics that could be used to organize a curriculum. The specific identifications of life, physical, or Earth and space sciences are not made at these levels, though the topics are usually based primarily in one of these areas. The disciplinary core ideas in the *Framework* were organized into K–2 and 3–5 grade-band endpoints, though the *NGSS* present them as grade-level standards. Students are introduced to all of the practices and crosscutting concepts starting in kindergarten and will experience these through a progression of sophistication, as with the core ideas and crosscutting concepts, building through to the end of high school. This is an important aspect, particularly for the practices. It is not the case that some categories of practices are more sophisticated than others. For example, "Using Mathematics and Computational Thinking" is not inherently more advanced than "Planning and Carrying Out Investigations;" both are done at all levels but vary depending on grade.

The elementary grades contain three Earth and space science standard topics, which each appear more than once: Space Systems (grades 1 and 5), Earth's Surface Systems (grades 2, 4, and 5), and Weather and Climate (grades K and 3). There is a greater integration of performance expectations at the elementary level. Performance expectations from those middle school Earth and space science topics not represented at the elementary level (History of Earth, Earth's Interior Systems, and Human Impacts) are incorporated into other standard topics in Earth and space sciences, life, or physical science. The goal is not to turn the 3–4 grade-level topics into broad umbrellas to cover all performance expectations, but rather the disciplinary core ideas for each module are interpreted so as to produce coherent storylines that provide sufficient depth into the content.

It is important that a well-connected progression of story lines continue through elementary and into middle school. For example, when Space Systems appears in grade 1, there is a focus on the patterns and cycles that students can observe in the sky. The patterns of the Sun, Moon, and stars provide observations by which students begin to understand foundational aspects of nature such as the passing of time, the cyclic aspect of some phenomena such as seasons, and the generation of light by stars. Students also encounter a significant role of engineering by recognizing that objects in the night sky can be seen in much greater detail with the help of telescopes. Students could record observations of natural phenomena (e.g., sunrise/sunset times) to identify patterns that allow for future predictions.

When Space Systems is revisited in grade 5, observational data concerning the day/night sky are expanded into more sophisticated practices. Students analyze data to develop an explanation for the relative brightness of the Sun and stars, providing the basis for understanding the scale of the universe. Students use observations to begin to explain the role of gravity in holding the Earth together. Students make the leap to developing models of the motions of the Sun, Moon, and Earth that will explain observations of repeating cycles of day and night, the length and direction of shadows, and the seasonal appearance of certain stars in the night sky. Corresponding engineering/technology concepts now investigate how telescope lenses operate. The story line now includes core ideas not only from the *Framework*'s big idea of Earth's Place in the Universe, but also from the physical science big ideas of Motion and Stability: Forces and Interactions and Waves and Their Interactions and from the engineering concept of Interdependence of Science, Engineering, and Technology.

## Middle school

Middle school Earth and space science standards are not broken out by grade but are presented in a 6–8 grade band in recognition of the wide array of curricular arrangements that may be practiced at this level. The six standard topics (Figure 1) represent roughly a year's worth of instruction. These could be taught in a single year but could also be broken out into a more integrated middle school curriculum. The performance expectations push students to more advanced levels, and the added complexity is addressed by the regular addition of "clarification statements" and "assessment boundaries," which help define their scope.

The middle school Earth and space science topics do not always align in a one-to-one manner with the big ideas from the *Framework*. This realignment allows for the performance expectations of the middle school topics to be presented in a structure that provides coherent story lines. For example, the story of Earth's history is best told by combining the history of plate motions and landform evolution with the paleontological record of biological evolution, which is why History of Earth contains *Framework* core ideas from both ESS1 and ESS2. Within the *Framework*, the subjects of natural resources and natural hazards are grouped together as part of the anthropospheric big idea ESS3.

However, in constructing story lines, the relevant parts of resources and hazards are better grouped with their appropriate topics. So, resources play an important role in Human Impacts, but the interior-related resources (e.g., minerals, energy sources) and hazards (e.g., earthquakes and volcanoes) appear in Earth's Interior Systems; the surface-related resources (e.g., soils, evaporates) and hazards (e.g., avalanches, landslides) appear in Earth's Surface Systems; and

**Figure 4. Earth and space science content sources**

These four literacy documents were used in constructing the Earth and space science content of the *Framework* (from Wysession 2011). These documents of essential concepts were produced by joint research and education community efforts that included the Earth Science Literacy Principles (ESLPs) (*www.earthscienceliteracy.org*), Essential Principles of Ocean Literacy (*www.coexploration.org/ocean literacy/documents/OceanLitChart.pdf*), Essential Principles and Fundamental Concepts for Atmospheric Science Literacy (*http://eo.ucar.edu/asl/index.html*), and Essential Principles of Climate Science (*http://cleanet.org/cln/index.html*). Of particular relevance were the ESLPs (Wysession et al. 2012), which were written with direct input from 350 Earth scientists and educators and reviewed by many hundreds more and consist of the nine "Big Ideas" and 75 "Supporting Concepts" that every citizen should know about Earth science.

weather-system hazards (e.g., tornadoes, floods, hurricanes) appear in Weather and Climate. Again, the *NGSS* are not a structured curriculum, and there are many different possible ways of organizing instruction around these performance expectations.

There is a clear progression from elementary school into middle school: The elementary school Space Systems lead into both Space Systems and History of Earth at the middle school level; the elementary school Earth's Surface Systems leads to both Earth's Interior Systems and Earth's Surface Systems, and the elementary school Weather and Climate becomes both Weather and Climate and Human Impacts at the middle school level.

Continuing with the example of Space Systems, the performance expectations in middle school take this topic to its natural progressions. For example, models of the Earth-Moon-Sun system are now used to construct explanations for more complex phenomena such as seasons, eclipses, and tides. The concept of gravity is now expanded from its role on Earth to address the motions of orbits within the solar system, and students are no longer just observing but are using models to explain the relative motions of the Sun, Moon, and Earth (seasons, lunar phases, eclipses). Students now examine the scale of both the solar system and galaxy. The contributions of engineering and technology to these topics are prominent in space and planetary science, particularly in acquiring the data that students analyze to examine the geologic processes of other objects in the solar system.

## High school

The *NGSS* present high school Earth and space sciences at a high level of complexity that progresses well beyond the middle school level. The structure of the U.S. secondary school science curriculum is largely based on recommendations made 120 years ago in a report by the

"Committee of Ten," which suggested that "physical geography" be taught in middle school, and that biology, physics, and chemistry be taught in high school (NEA 1893). Unfortunately for the Earth and space sciences, the emergence of modern geosciences in the 1960s and 1970s was not able to break into this construction, and the relegation of Earth and space sciences to middle school classrooms has only served to perpetuate the misconception that the geosciences are low-level sciences of categorization and classification.

This is no longer the case with the *NGSS*. For example, the high school Space Systems builds upon middle school Space Systems to include the physical science concepts of nuclear fusion and electromagnetic radiation in order to construct explanations for the big bang and formation of the universe, the creation of larger atomic nuclei during stellar nucleosynthesis processes, and the formation of the solar system based on analyses of observable data. Students mathematically and computationally apply Newtonian gravitational laws to the orbital motions of the solar system and analyze evidence to explain how changes in Earth's orbital parameters affect cyclic climate changes on Earth such as the repeating Ice Ages. This is not simplistic stuff.

Other high school Earth and space science standards also show a high level of sophistication and integration with other fields. The middle school Human Impacts progresses into the high school Human Sustainability, which provides students with complex, open-ended investigations that explore possible solutions to conflicts between increasing human populations, the development of dwindling resources, and the need to minimize the resulting impacts. Weather and Climate focuses on weather in middle school but on climate and climate change in high school. Climate change occurs over many time scales that span at least eight orders of magnitude (years to hundreds of millions of years) and include solar dynamics, the gravitational perturbations of other planets to Earth's orbital parameters, atmospheric and oceanic chemistry, radiation physics, oceanic and atmospheric circulation dynamics, biogeochemical cycles, volcanic eruptions, plate tectonics and mountain building, human activity, and the feedbacks between many of these. Student understanding of climate systems requires preexisting understandings in biology, chemistry, and physics.

## Model course maps

In fact, it turns out that most of the Earth and space science content needs to be taught following instruction in physics, chemistry, and biology, both for middle school and high school. An important part of the *NGSS* is a set of model course maps (Achieve Inc. 2013, Appendix J) that provide guidance on how to construct curricula around the *NGSS* performance expectations. Course Map 1, the most efficient in instruction and the most sensible in content progress, is a three-course integrated science curriculum for both middle and high school. It orders the science content by identifying the prerequisites needed for students to understand each core idea. The first year is largely physical science, with some Earth and space science; the second year is largely life science, with some physical and Earth and space science; the third year is mostly Earth and space science, with a large amount of life science as well. This holds for both middle and high school. This is similar to the "physics first" curriculum but with Earth and space science as the capstone.

The *NGSS* includes two additional model course maps. Course Map 2 would be the easiest way to align a curriculum with the *NGSS*: It consists of three courses each for middle and high school: Physical Science, Life Science, and Earth and Space Science. In theory, these could be taught in any order, but following the findings of Course Map 1, the ideal way to teach them would be in this exact order, again with Earth and Space Science as the capstone course for both middle and high school.

The last model, Course Map 3, is clearly the least acceptable; it tries to fit the Earth and space sciences into high school physics, chemistry, and biology courses. Because a significant amount of content would be taught out of order in terms of the needed progression of prerequisites, this model would be very inefficient as teachers would have to introduce concepts before they appear in the curriculum and then reteach them later. While most states now require three years of high school science, adding a fourth year would make Course Map 3 workable: a year of Earth and space science would follow courses in physics, chemistry, and biology.

Ironically, it is historically appropriate that Earth and space science should provide a natural capstone experience for the life and physical sciences. Nearly all sciences largely began in the form of Earth and space sciences through early attempts to explain the natural world. Sir Isaac Newton developed the theory of gravity (and calculus along with it) to explain the concept of objects orbiting Earth. Theories of magnetism began with studies of Earth's magnetic field. The field of chemistry began with alchemical studies of minerals and other Earth materials. Even theories of the history of life began with examinations of the paleontological rock record.

Earth and space sciences, in a modern curriculum, can provide students with concrete, understandable, and extremely societally relevant demonstrations of many of the concepts of biology, chemistry, and physics that are sometimes presented in those classes in more general and abstract ways.

## Challenges

The *NGSS* performance expectations, and Earth and space sciences in particular, will pose special challenges to their implementation. Most current science assessments test memorized facts. Shifting toward testing what students can do, instead, will require new approaches to assessment. Teachers may find many challenges: They may be unfamiliar with the aspects of engineering and technology newly integrated into the science curriculum; they may be unaccustomed to teaching from a practice-based approach as opposed to a concept-focused approach; and they may be unused to connecting science content with overarching crosscutting concepts, with aspects of the nature of science, and with Common Core concepts of math and English language arts. This will all require significant professional development.

The prominent role of the Earth and space sciences within the *NGSS* will also pose significant challenges to implementation. There aren't enough high school Earth and space science teachers in the United States to handle a significant increase in high school Earth and space science content, and a state's implementation of a curriculum based on the *NGSS* would need to provide for this. None of these challenges is insurmountable, however, and many efforts are underway to help address them. However they are ultimately organized, structured, taught, and assessed in all of the states that adopt them, the *NGSS* provide a remarkable opportunity to realize the long-identified potential for Earth and space sciences to be taught in exciting, engaging, encompassing, and relevant manners from kindergarten through twelfth grade.

*Michael E. Wysession* is associate professor of Earth and Planetary Sciences at Washington University in St. Louis, Missouri. He was the Earth and Space Science Design Team Leader for the NRC *Framework* and is on the Leadership Team for the writing of the *NGSS* with a focus on Earth and space science.

## References

Achieve Inc. 2013. *Next generation science standards. www.nextgenscience.org/next-generation-science-standards*

Ireton, F., D. W. Mogk, and C. A. Manduca, eds. 1996. Shaping the future of undergraduate earth science education: Innovation and change using an earth system approach. American Geophysical Union report. *http://serc.carleton.edu/shapingfuture/index.html*

National Center for Education Statistics (NCES). 2012. *The nation's report card: Science 2011.* Washington, DC: Institute of Education Sciences, U.S. Department of Education.

National Education Association (NEA). 1893. Report of the committee on secondary school studies (commonly known as The Committee of Ten Report). Washington, DC: Government Printing Office.

National Governors Association Center for Best Practices and Council of Chief State School Officers (NGAC and CCSSO). 2010. *Common core state standards.* Washington, DC: NGAC and CCSSO.

National Research Council (NRC). 1996. *National science education standards.* Washington, DC: National Academies Press.

National Research Council (NRC). 2007. *Taking science to school: Learning and teaching science in grades K–8.* Washington, DC: The National Academies Press.

National Research Council (NRC). 2008. *Ready, set, science.* Washington, DC: National Academies Press.

National Research Council (NRC). 2012. *A framework for K–12 science education: Practices, crosscutting concepts, and core ideas.* Washington, DC: National Academies Press.

Wysession, M. E. 2011. Implications for Earth and space in new K–12 science standards. *Eos* 93 (46): 465–466.

Wysession, M. E., N. LaDue, D. A. Budd, K. Campbell, M. Conklin, E. Kappel, G. Lewis, R. Raynolds, R. W. Ridky, R. M. Ross, J. Taber, B. Tewksbury, and P. Tuddenham. 2012. Developing and applying a set of Earth science literacy principles. *Journal of Geoscience Education* 60 (2): 95–99.

## Appendix

# A Look at the *Next Generation Science Standards*

*By Ted Willard*

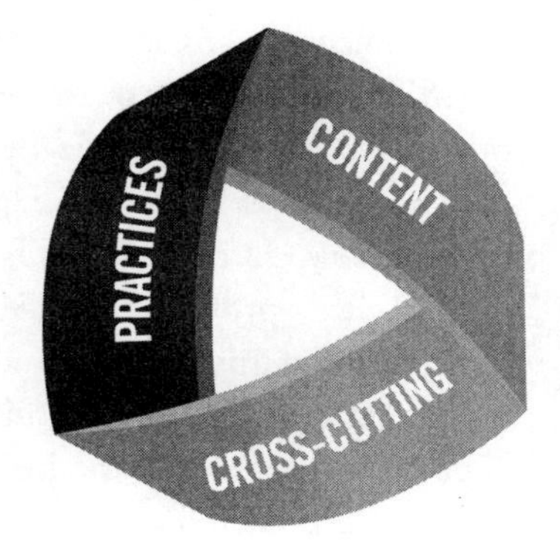


The following text and diagram provide an overview on the architecture of the standards.

## Overall architecture

The *NGSS* differ from prior science standards in that they integrate three dimensions (science and engineering practices, disciplinary core ideas, and crosscutting concepts) into a single performance expectation and have intentional connections between performance expectations. The system architecture of the *NGSS* highlights the performance expectations as well as each of the three integral dimensions and connections to other grade bands and subjects. The architecture involves a table with three main sections.

## What is assessed (performance expectations)

A performance expectation describes what students should be able to do at the end of instruction and incorporates a practice, a disciplinary core idea, and a crosscutting concept from the foundation box. Performance expectations are intended to guide the development of assessments. Groupings of performance expectations do not imply a preferred ordering for instruction—nor should all performance expectations under one topic necessarily be taught in one course. This section also contains *Assessment Boundary Statements* and *Clarification Statements* that are meant to render additional support and clarity to the performance expectations.

## Foundation box

The foundation box contains the learning goals that students should achieve. It is critical that science educators consider the foundation box an essential component when reading the *NGSS* and developing curricula. There are three main parts of the foundation box: science and engineering practices, disciplinary core ideas, and crosscutting concepts, all of which are derived from *A Framework for K–12 Science Education.* During instruction, teachers will need to have students use multiple practices to help students understand the core ideas. Most topical groupings of performance expectations emphasize only a few practices or crosscutting concepts; however, all are emphasized within a grade band. The foundation box also contains learning goals for connections to engineering, technology, and applications of science and connections to the nature of science.

## Connection box

The connection box identifies other topics in *NGSS* and in the *Common Core State Standards* (*CCSS*) that are relevant to the performance expectations in this topic. The "Connections to other DCIs in This Grade Level" contains the names of topics in other science disciplines that have corresponding disciplinary core ideas at the same grade level. The "Articulation of Disciplinary Core Ideas (DCIs) Across Grade Levels" contains the names of other science topics that either provide a foundation for student understanding of the core ideas in this standard (usually standards at prior grade levels) or build on the foundation provided by the core ideas in this standard (usually standards at subsequent grade levels). "Connections to the *Common Core State Standards*" contains the coding and names of *CCSS* in mathematics and in English language arts that align to the performance expectations.

*Ted Willard* is a program director at NSTA.

# Inside the NGSS Box

## What is Assessed

A collection of several performance expectations describing what students should be able to do to master this standard.

## Foundation Box

The practices, core disciplinary ideas, and crosscutting concepts from *A Framework for K–12 Science Education* that were used to form the performance expectations.

## Connection Box

Other standards in the *Next Generation Science Standards* or in the *Common Core State Standards* that are related to this standard.

*Based on the January 2013 Draft of NGSS*

### Title and Code

The titles of standard pages are not necessarily unique and may be reused at several different grade levels. The code, however, is a unique identifier for each set based on the grade level, content area, and topic it addresses.

3-PS2 Motion and Stability: Forces and Interactions

3-PS2-a. Carry out investigations of the motion of objects to predict the effect of forces on an object in terms of balanced forces that do not change motion and unbalanced forces that change motion.

3-PS2-b. Investigate the motion of objects to determine when a consistent pattern can be observed and used to predict future motions in the system.

3-PS2-c. Investigate the effect of electric and magnetic forces between objects not in contact with each other and use the observations to describe their relationships.

3-PS2-d. Apply scientific knowledge to design and refine solutions to a problem by using the properties of magnets and the forces between them.

Science and Engineering Practices

Disciplinary Core Ideas

Crosscutting Concepts

### Codes for Performance Expectations

Codes designate the relevant performance expectation for an item in the foundation box and connection box. In the connections to common core, italics indicate a potential connection rather than a required prerequisite connection.

### Performance Expectations

A statement that combines practices, core ideas, and crosscutting concepts together to describe how students can show what they have learned.

### Clarification Statement

A statement that supplies examples or additional clarification to the performance expectation.

### Assessment Boundary

A statement that provides guidance about the scope of the performance expectation at a particular grade level.

### Engineering Connection (*)

An asterisk indicates an engineering connection in the practice, core idea, or crosscutting concept that supports the performance expectation.

### Scientific and Engineering Practices

Activities that scientists and engineers engage in to either understand the world or solve a problem.

### Disciplinary Core Ideas

Concepts in science and engineering that have broad importance within and across disciplines as well as relevance to people's lives.

### Crosscutting Concepts

Ideas, such as *Patterns* and *Cause and Effect*, which are not specific to any one discipline but cut across them all.

### Connections to Engineering, Technology, and Applications of Science

These connections are drawn from the disciplinary core ideas for engineering, technology, and applications of science in the *Framework*.

### Connections to Nature of Science

Connections are listed in either the practices or the crosscutting connections section of the foundation box.

# Index

E